Looking at the stars at night, we have all wondered where we have come from and why we are here. *Our Cosmic Origins* traces the remarkable story of the emergence of life and intelligence right through the complex evolutionary history of the Universe. Starting from the searingly hot soup of exotic particles in the Big Bang, we are taken on a breathtaking chronological tour, right up to the present day, and the recent findings of bacteria on Martian meteorites and planetary systems around other nearby stars.

Armand Delsemme weaves a rich tapestry of science, bringing together cosmology, astronomy, geology, biochemistry and biology in this wide-ranging book. In following the complex chronological story, we discover how the first elements formed in the early Universe, how stars and planets were born, how the first bacteria evolved towards a plethora of plants and animals, and how the coupling of the eye and brain led to the development of self-awareness and, ultimately, intelligence.

Throughout the book, Professor Delsemme shows how this ascent always seems to follow the easiest route. He suggests that, however complex and tortuous this evolution has been, it is unlikely to be unique. The author concludes with the tantalizing suggestion that the existence of alien life and intelligence is likely, and examines our chances of contacting it. This provocative book provides the general reader with an accessible and wide-ranging account of how life evolved on Earth and how likely it is to exist elsewhere in the Universe.

OUR COSMIC ORIGINS

OUR COSMIC ORIGINS

From the Big Bang
to the emergence of
life and intelligence

ARMAND DELSEMME

Professor of Astrophysics
Department of Physics and Astronomy
University of Toledo, Ohio

 CAMBRIDGE
UNIVERSITY PRESS

PUBLISHED BY THE PRESS SYNDICATE OF THE UNIVERSITY OF CAMBRIDGE
The Pitt Building, Trumpington Street, Cambridge CB2 1RP, United Kingdom

CAMBRIDGE UNIVERSITY PRESS
The Edinburgh Building, Cambridge CB2 2RU, United Kingdom
40 West 20th Street, New York, NY 10011-4211, USA
10 Stamford Road, Oakleigh, Melbourne 3166, Australia

First published in French as *Les Origines Cosmiques de la Vie* by Flammarion in the *Nouvelle Collection Scientifique* 1994

First published in English by Cambridge University Press 1998
as *Our Cosmic Origins: from the Big Bang to the emergence of life and intelligence*

The English edition has been revised, updated and expanded

Printed in the United Kingdom at the University Press, Cambridge

Typeset in Monotype Sabon 11/13 pt   [SE]

A catalogue record for this book is available from the British Library

ISBN 0 521 62038 4 hardback

To Delphine

'to be in love is not to look at each other,
but to look together in the same direction.'

Saint-Exupéry, *Terre des Hommes,* 1939

CONTENTS

CONTENTS

FOREWORD

From the Big Bang to the human brain, the Universe has gone some 15 billion years of cosmic, physical, chemical and biological evolution. Now, our own little speck of dust is beginning to look into itself and ponder its origin, nature and significance.

How did it all happen? What is known, suspected, or assumed about each of the steps whereby time and matter first arose out of nothing, elementary particles condensed out of the original plasma, and out of them, in turn, the atoms of the various elements came to be? What is known about how galaxies were born, spawning billions of stars, many probably surrounded by planetary systems? What is known about how, on one particular planet, which happened to combine a special set of physical conditions, life emerged and evolved, finally producing conscious, thinking beings?

How much of this extraordinary history is due to deterministic forces, how much to chance? Did it happen only once? Or does the cosmos contain many planets that have given rise to life, perhaps even to intelligent life? What is it about the cosmological constants that endows our Universe with its unique properties? Is only one such universe possible? Or are there many universes, of which ours happens to bear life and intelligence, and thus is knowable, because of a special combination of cosmological conditions? What triggered the Big Bang? Was it a creative act of God or just randomly fluctuating nothingness?

These questions have become popular themes for contemporary thinkers, from cosmologists to theologians. Rarely have they been addressed by a single author. Armand Delsemme has risen to the challenge. As an expert on comets, some of which may have brought to our young planet the water and organic chemicals out of which life arose 4 billion years ago, he is well placed to tell us about what preceded and what followed that decisive event. He has done his homework well, using the same even-handed and well-informed style to describe all the phases of cosmic history. He explains the facts as they are known and the hypotheses as they have been propounded, in a sober and objective manner that seldom allows whatever personal biases he may harbor to get through.

On the whole, this is an eminently readable and informative account, consistently written in a way that tries to eschew technical difficulties, while remaining solidly anchored in the realities of scientific concepts. Readers could not wish for a better introduction to the history of the Universe, or a better appetizer for more solid fare on any one period of this astonishing history.

Christian de Duve
Nobel laureate

PREFACE

Humankind has just awoken to the cosmic origins of the human adventure and I have striven to reconstruct this story as science now sees it. In order to keep it accessible to the enquiring reader, I have not used mathematics but followed the thread of chronology.

Recently, it has become possible to tell this story without leaving too many gaps. At the beginning of this century, the different natural sciences, originating in the same discipline of natural philosophy, had become specialized; they were cut off from one another. Now, physics, cosmology, astronomy, chemistry, geology, paleontology, biochemistry and biology are coming together again.

So my thread follows the Universe from the moment of its beginning; I tell how its expansion led to the formation of the galaxies and stars. The pursuit of the thread shows us how life could finally emerge, through the most probable cosmic processes, while using the most abundant elements made in the crucible of the stars' cores.

I have tried to keep the story simple; I cannot say that I have always succeeded. Occasionally, I had to cut short explanations that might have become tedious. The pages that seem too difficult can be skipped; in order to help the reader to follow the thread to the final goal, I have tried to give short summaries throughout the book.

The purpose of the first chapter is to familiarize the reader with the extremely large and the extremely small. The physics of elementary particles is needed to understand the Big Bang; the reader who wishes to avoid the technicalities can skip a couple of pages and go to Chapter 2, where the chronology of the early events is laid out.

In a popular book like this one, it is not possible to debate all doubts and explore alternative views. Nothing is ever certain in science, since its progress is steadily reached by trial and error. This book is only a balance sheet for the end of the twentieth century; when in doubt, I have sometimes accepted the interpretation that looked most reasonable to me. Some scientists who read this may discover that I have glossed over details of their particular discipline. That is the price of my effort that aims at a comprehensive view of the human condition in a still very mysterious Universe, at a time when it has become impossible to be an expert in all areas of science. I hope therefore that, in my striving for an integrated account, imperfections will be forgiven. My major goal has been to show that it has become possible to assemble enough of the story of our cosmic origins that we can follow it without any major gaps in our understanding.

An interdisciplinary approach like this often meets another obstacle: jargons from different scientific disciplines are difficult to assimilate. I have cut their use as much as possible, and when a paraphrase would be too lengthy I have explained terms in a Glossary at the end of the book.

Questions still unresolved and related philosophical issues have been left for the final chapter of the book. Finally, in order to express the extremely large or small numbers encountered in astronomy or in physics, I have used exponential notation, assuming that the reader knows that $10^3 = 1000$, $10^6 = 1$ million, $10^9 = 1$ billion and more generally, that 10^n indicates that 1 is followed by n zeros. Similarly, for the negative exponent, e.g. 10^{-3}, 1 is *divided* by 10^3.

Delphine Delsemme, my wife, has been of great help, not only in reviewing the final text, but also in discussing ideas and clarifying difficult passages. Without her assistance and her constant support, I do not know whether I would have had the patience to finish the task.

Revisions incorporated into the English edition have been considerably helped by comments and constructive criticisms of many friends; André Brack, Christian de Duve, Gérard de Vaucouleurs, Audouin Dollfus, Léo Houziaux, Henri Mertens, André Lausberg, Jean-Claude Pecker, Hubert Reeves, Evry Schatzman and John Schlag, who read the first edition in French. Phil Mange did even more: in his enthusiasm to push me to publish the book in English, he even produced a preliminary translation from the French.

The English edition was updated in 1997, to include in particular the recent discoveries of giant planets around nearby stars, and the possible identification of fossil bacteria inside meteorites coming from Mars. Both discoveries must be carefully interpreted before any conclusions are drawn. However, they are in line with the thread of the book, and both were foreseen in the 1994 French edition. Even if new discoveries are in line with my expectations, they are important enough to be subjected to careful scrutiny.

Finally, I want to thank the whole team at Cambridge University Press, particularly Adam Black and Sandi Irvine, for their invaluable help.

ACKNOWLEDGMENTS

FIGURE 2.1: Adapted from data given in Audouze, J. and Vauclair S. 1986 *An Introduction to Nuclear Astrophysics*, D. Reidel Publ. London.

FIGURE 2.2: From data from the COBE satellite, NASA, 1992.

FIGURE 2.3: NASA/Goddard Space Flight Center, Greenbelt, MD 20771, USA.

FIGURE 2.4: Reproduced with permission from Seldner, M., Siebers, B. L., Groth, E. J. and Peebles, P. J. E. 1977 *Astronomical Journal*, **82**, 249.

FIGURE 2.5: Courtesy of M. Geller and J. Huchra, Smithsonian Astrophysical Observatory; originally published in Geller, M. and Huchra, J. 1989 *Science*, **246**, 897.

FIGURE 2.6: Courtesy of Hale Observatories, archival plate.

FIGURE 2.7: L. Ferrarese (Johns Hopkins University) and NASA.

FIGURE 3.2: Richard P. Boyle, S. J., Vatican Observatory.

FIGURE 3.3. Kitt Peak National Observatory; Cerro Tolo Inter-American Observatory.

FIGURE 3.5: Adapted from Chiosi, C. and Maeder, A. 1986 *Annual Review of Astronomy and Astrophysics*, **24**, 329.

FIGURE 4.1: Courtesy of D. B. Hoisington, adapted from data given in Hoisington, D. B. 1959 *Nucleonics Fundamentals*, McGraw-Hill Book Co., New York.

FIGURE 4.2: Drawn from cosmic abundances of elements given in Allen, C. W. 1976 *Astrophysical Quantities*, Athlone Press, London.

FIGURE 4.3: C. R. O'Dell/Space Telescope Science Institute.

FIGURE 4.6: Courtesy of the European Space Agency.

FIGURE 4.7: Courtesy of ESA (Giotto) and Uwe Keller.

FIGURE 5.1: Reproduced from M. Loewy and P. Puiseux, *Atlas photographique de la Lune*, 1912.

FIGURE 5.3: Reproduced with permission from Schopf, W. 1994 *Early Life on Earth*, Columbia University Press, New York.

FIGURE 5.8: Adapted from Lehninger, A. L. 1978 *Scientific American*, March, 121.

FIGURE 5.9: Courtesy of C. Woese and H. Noeller, University of Santa Cruz; originally published in Noeller, H. F. and Woese, C. R. 1981 *Science*, **212**, 403.

FIGURE 5.11: Courtesy of Sherwood Chang, 1982, adapted from lecture notes.

FIGURE 6.3: Copyright C. A. Henley/BIOFOTOS.

FIGURE 6.4: Adapted from Emiliani, C. 1992 *Planet Earth*, Cambridge University Press, Cambridge: redrawn and put to uniform scale from different drawings.

FIGURE 6.6: Reproduced with permission from Emiliani, C. 1992 *Planet Earth*, Cambridge University Press, Cambridge.

FIGURE 8.2: Reproduced with permission from Briggs, G. A. and Taylor, F. W. 1982 *The Cambridge Photographic Atlas of the Planets*, Cambridge University Press, Cambridge.

FIGURE 8.3: Courtesy of NASA, 1996.

FIGURE 8.4: Courtesy of NASA, Galileo spacecraft, 1990–1992.

LOCATING HUMANS IN THE UNIVERSE

'For what is Man in nature? A nothingness in respect to
infinity, a whole in respect to nothingness, a median
between nothing and everything'

Blaise Pascal, *Les Pensées*, 1670

Introduction

Even more than the Renaissance period, the twentieth century will be
remembered in human memory as an extraordinary era in every
regard. The awareness of our true position, and of our isolation, in an
immense and mysterious Universe began nearly 400 years ago, but
recently it has expanded enormously (see Figure 1.1A and B).

At the end of the nineteenth century, we did not know where we
were, or where we came from; we did not even realize that we did not
know it. The vastness of space and time had always been thought to be
beyond any possible observation or experiment; in a word, their study
was considered to be a part of *metaphysics*. Metaphysics concerns
everything that *might* exist, but which we have no means of detecting.
In contrast, the physical world is made up of what we can see, touch,
hear, taste and smell, i.e. observe. In this sense it can be said that *angels*
are a part of metaphysics, whereas a *chair* is part of the physical world.

Over the last 300 years, we have invented new means of detection
that have extended our senses and give them 'feelers'. We have rolled
back the limits of metaphysics more and more. Our detectors have
recognized a number of invisible waves in the domain of the physical world. Although not seen, heard or felt, these waves have
become so familiar that they are now part of our daily life: radio,

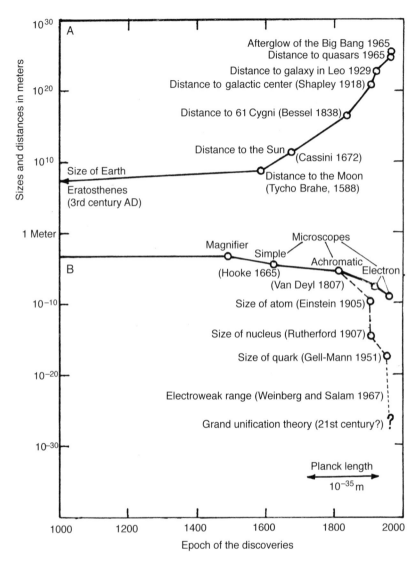

FIGURE 1.1   Exploration of (A) the extremely large and (B) the extremely small during the last millennium.

television, X-rays, gamma-rays, infrared and ultraviolet rays. We have probed the extremely distant as well as the extremely small with techniques that could not have been imagined three centuries ago.

Yet it would be naïve to believe that metaphysics is about to disappear. We still do not know nearly enough to guess at what is hidden by what is presently unknown. However, it appears that our inquiries have made sufficient progress to begin to perceive limits. These limits seem to exist in time as well as in space, in the extremely small as well as in the extremely large; they bring problems that we are not yet able to comprehend. For example, the extremely large is perhaps not infinite, in spite of being without limits. The infinitely small might not exist, since the extremely small is governed by strange laws which prevent us from going any further because they seem to defy common sense. Moreover, eternity does not seem to have existed in the past, because there was a 'grand beginning' before which there probably was no 'before'. Time could also come to an end in the distant future.

Investigation of the distant past has led to the understanding that the Universe has not stopped evolving towards a growing complexity. This is a fundamental discovery that illuminates the human condition in a new light. From a beginning of extreme simplicity, the Universe has proliferated more and more complex structures, of which we are the outcome. The primordial explosion first produced elementary particles, then assembled them into light atoms, which formed the first stars. These stars cooked a great variety of heavy atoms in their cores, and then exploded, scattering the elements deep into interstellar space. These new atoms combined into the first grains of solid matter and the first organic molecules. Solid matter allowed the birth of planets such as the Earth; on such a planet, organic molecules brought all the chemicals needed for the emergence of life. Seen from this perspective, the growth in complexity of biological structures seems to follow the inevitable pathway of preexisting cosmic processes toward an ever growing sophistication. The biological evolution that so far has led to intelligence seems to be the consequence of those cosmic processes that started from the onset of a slight asymmetry.

Together we are going to try to reconstruct, step by step, the story of the Universe. It will lead us from the Big Bang, to the shaping of galaxies, then the condensation of stars in the midst of galaxies, to the cooking of most of the chemical elements in the heart of stars, and to their dispersal in interstellar space. I will describe the birth of the second-generation stars, which for the first time allowed the formation of rocky planets such as the Earth, and I will sketch the complex formation of our Solar System. We will discover the crucial role of the giant planets, which have deflected cometary orbits, so that comets will carry to Earth the organic molecules needed for the emergence of life. We will follow the long process of biological evolution on the Earth, and examine the ways and means by which astronomical and meteorological phenomena have shaped the incredible complex branching and pruning of the evolution of species. Thus we will come to understand how biological evolution and natural selection have finally resulted in human beings.

Needless to say, this history of the Universe still has many gaps. I will have to paraphrase and simplify, restricting myself to sketching only the main features. In spite of all these gaps, most known pieces of the puzzle seem to fall in place, so that we can make a reasonable guess at the whole picture, and the results seem extraordinarily promising. Before embarking on a chronology of the events in the whole Universe, it may be useful to recall the sizes and distances of the objects and structures that form the observable cosmos.

Let us first recall that, in spite of the great speed of light, the distances to be considered are so enormous that, when we look into them, we look into the past. On Earth, this past is indistinguishable from the present; at the speed of 300 000 km/second, light from a flash of lightning beyond the horizon reaches us in less than 1/10 000 second. But light already takes more than 1 second to reach us from the Moon, 8 minutes to come from the Sun, and 4 years from the nearest star. These times become incomprehensible for objects outside our Galaxy.

The light we see from the Andromeda Galaxy (barely visible to the naked eye) left there 2.5 million years ago; that is, long before the emergence of civilizations on Earth. The light from distant quasars left more than 10 billion years ago – a very long time before our Earth even existed. We do not see these objects as they are at present, but as they were at different times in the past.

I will begin this chapter by describing the general perception of the world three centuries ago. Then I will briefly explore, in measured steps, the depths of space. Finally, I will proceed in the same way into the microcosm revealed by present-day physics. This approach is needed if we are to understand the connections between elementary particles and cosmology at the beginning of the Universe.

Pascal's two infinities

Quantitative exploration of the Universe had barely started before the seventeenth century. In 1617, Johannes Kepler discovered that the square of the periods of the planets' revolutions around the Sun are proportional to the cube of their distances – known as Kepler's Third Law. Since Saturn takes about 30 years to make a circular journey around the sky before coming back to the same constellation, it was easy to compute that $30^2 = 900$, and conclude that the distance of Saturn from the Sun would have to be nearly ten times larger than that of the Earth (because $10^3 = 1000$). This enormous distance was difficult to accept, because it shattered antiquity's concept of crystalline spheres in the sky.

A little later, in 1672, the astronomer Jean-Dominique Cassini, who came from Italy to France, at the request of King Louis XIV, to start the Observatoire de Paris, was able to determine the scale of the Solar System through careful measurement of the distance from the Earth to the Sun. Dubbed the astronomical unit (or AU in short), this distance is approximately 150 million kilometers.

During the same period, Dutch naturalists were for the first time exploring the world of cells and microbes. Using the single-lens microscope (whose objective was a tiny glass ball) they could distinguish details of less than a hundredth of a millimeter. Thus, Jan Swammerdam discovered in 1660 the existence of red blood cells, while in 1674 Antonie van Leeuwenhoek discovered infusoria, bacteria and spermatozoa.

The educated people of the time were astonished by these discoveries and wondered what was hidden in the infinitely small and the infinitely large. Shortly before his death in 1662, the French scientist and philosopher Blaise Pascal wrote of this in his *Pensées*, published posthumously in 1670. In order to better appreciate the philosophical 'dizziness' arising from the human condition, he emphasized the 'disproportionality' of humankind in comparison with the two infinities, the infinitely large and the infinitely small. He first meditated on the immensity of the Solar System, just revealed at the time by Kepler's

Third Law. Although he did not know any stellar distance, he wanted us to feel the immensity of the depths of space:

> However hard we try to inflate our conceptions beyond imaginable realms, we give birth to atoms only at the expense of real things.

The discovery of the microscopic world by the Dutch naturalists amazed Pascal just as equally, since he insisted on the fact that the infinitely small hides 'another marvel just as astonishing'; he wished us to contemplate 'a new abyss' where we could see 'an infinity of universes, each with its firmament, its planets, its Earth . . .; on this Earth, animals . . .'. Pascal's grandiose vision still seemed prophetic 200 years later. In 1860 we still had no idea of the size of the atoms, and we had just attempted to measure the distance to some of the nearest stars.

The exploration of the extremely large and the extremely small began in earnest only in the twentieth century, and its results have been even more surprising than anything that the author of *Les Pensées* would have been able to dream of. First, the two infinities are in no way similar, because no structure repeats itself at the different scales. The atom has nothing to do with a Solar System in miniature, in contrast to its first primitive model, proposed by the Danish physicist Niels Bohr in 1913, but discarded in 1926 in favor of quantum mechanics. Of course, the natural laws remain valid for all scales, but their influence depends essentially upon the scale under consideration. For instance, gravity remains valid at the atomic level, but it is too weak to work effectively. The electromagnetic force prevails at the atomic scale, whereas at the still much smaller scale of the atomic nucleus, much stronger nuclear forces prevail.

For Pascal, the greatest surprise would probably have been that natural limits have been found for the extremely large as well as for the extremely small; we are not yet sure of the meaning of these limits. In the extremely large, there is the possibility (but as yet no certainty) that the Universe is not infinite, but could be without limits because the geometric structure of space–time would close it back upon itself.

And what about the 'Big Bang', the explosion that is thought to have started the whole Universe? The theory of the Big Bang clarifies and explains the cause of a number of observational results that would otherwise remain mysteries.

The existence of the Big Bang introduces a natural limit, not only for the time elapsed, but also for the volume of the Universe accessible to observation, because of the finite speed of propagation of light. At that distant time in the past, the enormous heat of the Big Bang made all space as bright as the surface of our Sun. It is the afterglow of this light, shifted to millimeter wavelengths by the expansion of the Universe, that we detect in the depths of space. In a way, it is an impenetrable veil that hides from us the first moments of the Universe.

On the other hand, for the very small, the study of the world of quanta has revealed a haziness inherent in the nature of things; it arises from the discrete structure, both wavelike and corpuscular, of all particles of matter and of radiation. This fundamental haziness implies in particular that, in order to detect smaller and smaller particles, we must bombard them with other particles moving at higher and higher speeds. Among physicists, this has spurred a rush to create larger and larger particle-accelerators, operating at higher and higher energies (and with more prohibitive budgets!).

Thus, we now glimpse a perhaps finite Universe, but more extended than anything Pascal would ever have dreamed of, made up of elementary particles smaller than any he would have been able to imagine, but perhaps still limited by the structure of space itself, which we are only beginning to try to understand.

In the following sections, since we must feel comfortable with imagining structures of prodigiously different dimensions, I am going to give a brief summary of the scale of objects in the Universe. To avoid the bewilderment caused by large numbers, I will move up or down by reference steps. But the steps themselves must not be too big; whenever possible, I choose a factor close to 10 000. This factor represents reasonably well the limit of our perceptions without the help of any instruments. Indeed, 10 000 steps take us just to the horizon, while a ten-thousandth of a meter (0.1 mm) is the ordinary limit of the resolving power of the human eye.

For our ascent (or descent) there is fortunately a hierarchy of structures to help us.

Ascent to the extremely large

First step: the mountains (10^4 m = 10 km)

The meter is appropriate to the size of human beings; 10 000 m (10 km) takes us beyond the horizon or, if we travel upward, a little higher than the highest mountains. Ascending a very high mountain already gives us the feeling of the smallness of humans and of our insignificance with regard to nature. And thus I have just described the scope of primitive peoples. The nomadic hunter living on a hunting territory (some scores of square kilometers) could not comprehend the size of the Earth – the next step.

Second step: the planets (10^8 m = 100 000 km)

The size of the Earth was successfully assessed for the first time by Eratosthenes of Alexandria in the third century AD. He measured the shadow cast by a vertical stick in Alexandria, at the moment when he knew that the Sun was vertically overhead at Aswan in Upper Egypt. He concluded that the angle separating the two verticals was 1/50 of the circumference. Since the two cities were 5000 stadia apart, he deduced that the circumference of the Earth measured $50 \times 5000 = 250\,000$ stadia. The stadium was the field where the Greeks practiced sports; examination of the ruins of ancient Greece shows that its length was approximately 160–170 m. Eratosthenes' result was close to the correct value, since $250\,000 \times 160$ m = 40 000 km, the Earth's circumference. With its diameter of about 13 000 km, the Earth is a very modest planet. A giant planet like Jupiter has a diameter 11 times as large (140 000 km).

Third step: the giant stars (10^{12} m = 6.7 AU)

When he defined the astronomical unit (AU) as the mean distance from Earth to Sun, Cassini also estimated the diameter of the Sun: about 1 400 000 km or 0.009 AU. The luminosity, distance and diameter of the *red giant* stars was determined only in the twentieth century, by the use of the Cepheid stars. Some of the red giants have diameters larger than the distance from the Earth to the Sun. Betelgeuse, for instance, a red giant in the Orion constellation, has a diameter of 4 AU. By and large, red giants are stars at the end of their evolution. When they have transformed most of their core's hydrogen into helium, the core collapses upon itself and rises to a much higher temperature. Radiation from the central core becomes so intense that it blows out and distends the outer layers of the star to the enormous observed diameter.

Fourth step: the Solar System (10^{16} m = 1 light-year)

The light-year (l.y.) is the distance travelled by light in one calendar year, at a constant speed of 300 000 km/second; it is equal to 63 000 AU. The planetary orbits are all concentrated within a tiny zone of 50 AU radius around our Sun, but the attraction of the Sun dominates to a distance at least 1000 times larger, as shown by the 100 billion comets present in the Oort cloud. Although not detectable from Earth, the presence of the Oort cloud is demonstrated each year by two or three 'new' comets detached from it by galactic tides. The existence of the Oort cloud surrounding the Sun fixes the size of the Solar System at more than 1 l.y.

Distances to neighboring stars are also included in the fourth step, because they are only a few light-years away. The German astronomer Friedrich Bessel determined the first distance to a star in 1838, by measuring the parallax of 61 Cygni. Parallax is the angle through which the star appears to oscillate over one year against a background of distant reference stars, because of the yearly motion of the Earth around the Sun. Bessel found that 61 Cygni oscillates around its mean position by 0.30″ (3/10 of a second of arc) which corresponds to a distance of 11 l.y. Comparison of a few bright stars showed later that the parallax of Proxima Centauri was the largest, so that this star was the nearest to the Solar System. Since the parallax of a star diminishes as its distance increases, parallaxes can be measured, but with less and less precision, up to distances of about 1000 l.y., thanks in particular to the recent observations by the European satellite Hipparcos.

Fifth step: the galaxies (10^{20} m = 10 000 l.y.)

In 1610, when Galileo turned his first astronomical telescope toward the Milky Way, he discovered to his amazement that this milky ribbon, barely visible across the sky on moonless nights, is actually a collection of stars too faint to be seen individually with the naked eye. In 1780, the Englishman William Herschel proposed that the Milky Way is actually a disk composed of millions of stars. We can see this disk from within because our Solar System is inside and part of it.

In 1918, the American Harlow Shapley determined that our distance from the center of this enormous system of some 600 billion stars was about 25 000 to 30 000 l.y. It was only after the Second World War that radioastronomy finally detected the spiral structure of the immense arms of interstellar hydrogen which are part of the Milky Way. This demonstrated its similarity to numerous spiral-form galaxies seen in all directions in the depths of space.

The Milky Way is our Galaxy. It is a giant galaxy, ten times as large as the average. The size of our fifth step coincides with the average diameter of the multitude of small galaxies.

Sixth step: the clusters of galaxies (10^{24} m = 100 million l.y.)

Far out in the depths of space, we see that the galaxies are not uniformly distributed. They spread out in irregular groups separated by enormous distances of totally black space. In particular, our Galaxy is part of a group of some 30 galaxies; this is the so-called *Local Group*. The two giant galaxies of the group are ours (diameter 100 000 l.y.) and that of Andromeda (130 000 l.y.), which is 2.5 million l.y. from us. There are also four or five galaxies of average size, like M33 in the Triangle (diameter 60 000 l.y.), the two Magellanic clouds (30 000 and 15 000 l.y.) plus a good score of small galaxies (all less than 10 000 l.y.). These 30 or so galaxies are all contained in a quasi-spherical volume of the order of 4 or 5 million l.y. in diameter. Our Galaxy is near the periphery, while the Andromeda Galaxy is at the center. Variable giant stars like the Cepheids as well as novas are still distinguishable individually in the Local Group, and this permits measurements of sizes and distances in the group.

Our Local Group is a very small cluster of galaxies. The richest of the next clusters of galaxies is the Virgo Cluster, consisting of at least 2500 bright galaxies. Within it, the Hubble Space Telescope (HST) detected in 1992 numerous bright blue individual stars, confirming that its distance is close to 50 million l.y. This was an important measurement, because it served to calibrate larger distances and, from these, the expansion rate of the Universe. The distance of the Virgo Cluster corresponds, within a factor of 2, to our sixth step.

Seventh step: the visible Universe (1.5×10^{26} m = 15 billion l.y.)

Our last step is the least extended, because it is only 150 times larger than the previous one, and that is because we cannot see any further. This is not for lack of more powerful instruments, but because we have reached a natural limit called 'the event horizon'. It occurs because objects that are more and more distant are moving away from us faster and faster; therefore their light is shifted towards longer and longer wavelengths by their speed of recession. For this reason, the energy of the luminous photons diminishes, and it would become nil for galaxies escaping at the speed of light. The most distant objects still visible in our telescopes move away with a speed 93% of that of light. At that distance, quasars (which I will define later) are much more numerous than galaxies. Quasars are practically all found at distances corresponding to speeds of recession of between 85% and 93% of the speed of light. Finally, from the deepest space, the afterglow of the Big Bang comes from all directions. Its light has been shifted into millimeter waves; hence its recession speed is beyond 99.99% of the speed of light. Thus it comes almost from the event horizon; that is, from the limit of the immensity of time and space that is still directly perceptible.

Descent to the extremely small

In the descent to the extremely small, I will use as reference *the cell*, which is the basic element of life forms; *the atom*, which is the building block of the molecule; and *the quark*, which is the building stone of the atomic nucleus.

First step: the cell (10^{-6} m = 1 micron = 1 μm)

The micron (which some prefer to call the micrometer, to respect the formal rules of the metric system) is equal to one-thousandth of a millimeter. Under normal conditions, the eye can discern structures as small as 100 μm. The cells of animals and plants usually fall between 10 and 100 μm, diameter thus are only seen with a very good magnifier, or better with a microscope. The internal structure of these cells is very complex. They have a well-differentiated central nucleus, which holds the genetic recipe for making similar cells, but they also possess 'organelles', small organs with well-defined functions. For example, in animals, the *mitochondria* release energy through oxidation; in plants, *chloroplasts* use photosynthesis for the same purpose.

The cells of *blue-green algae* and *bacteria* are smaller; they range in size from 10 to only 1 μm. Although much less complex and devoid of nucleus and organelles, they still total from 1 to 100 trillion atoms (10^{12} to 10^{14}). *Viruses* are of the order of one-tenth of a micrometer. They are not cells but parasites that could not survive solely by themselves. However, they use nearly the same genetic code as all other forms of life; hence to survive, they shelter themselves within living cells that provide these 'parasites' with food and allow them to multiply. Viruses consist of only about 10 billion atoms (10^{10}).

Second step: the atom (10^{-10} m = 10^{-4} μm = 1 Å)

The lightest of the atoms is hydrogen. It has a diameter of 1 Å (ångström unit) or, to be precise, of 1.06 Å. The ångström is 1/10 000th of a μm. It is a unit better adapted than the nanometer (= 10 Å) to describe atomic dimensions.

All atoms of hydrogen are absolutely identical and indiscernible from one another. Their mass is concentrated in a central core still 10 000 times smaller: the atomic nucleus. The nucleus is charged with positive electricity, neutralized by an electron carrying exactly the same charge of negative electricity. In order to form a stable atom, the electron, which is 2000 times lighter than the central nucleus, revolves around it at a speed of about 2000 km/second, hence it makes 30 billion orbits each millionth of a second. However, its orbit is not stable in the sense of a planetary orbit around the Sun. Actually, it wobbles within a whole spherical volume that defines the diameter of the atom. It is this electronic cloud or shell which also defines the minimal distance separating neighboring atoms, since the shells cannot interpenetrate.

The atoms of other elements are barely larger than those of hydrogen. For example, the oxygen atom has a diameter of 1.10 Å. We must move to elements a great deal heavier for the difference to become substantial. The iron atom has a diameter of 3.2 Å, and that of gold 3.6 Å, even though their respective weights are 56 and 197 times that of the hydrogen atom.

At the atomic level, we meet for the first time a property completely foreign to the world of familiar objects. Particles (such as the electron or the atomic nucleus) behave like waves (which can be added or subtracted depending on their phase), while waves (such as light or X-rays) behave like particles (as if they were little bundles of energy, well separated and positioned in space). As Louis de Broglie demonstrated in his doctoral thesis in 1924, there is no fundamental difference between waves and particles. However, a large part of the difficulty in comprehending phenomena on the atomic scale arises from that fact. For instance, if the electron cannot be located on a stable orbit around the nucleus, that is precisely because it is also a wave whose vibrations are stationary and produce the spherical cloud that surrounds the atomic nucleus.

As for molecules, they are aggregates of a certain number of atoms, linked together by forces that result from the oscillating nature of the haze of surrounding electrons. The size of atoms remained a mystery up to 1905, when Einstein showed that it was possible to explain the Brownian motion of microscopic pollen grains by their collisions with the molecules of water in which they were suspended. Einstein concluded that the atoms were 10 000 times as small as the pollen grains. By 1912, a dozen different techniques to measure the size of atoms had been discovered by other physicists. The good agreement between these results, masterfully demonstrated by the French physicist Jean Perrin, proved for the first time that the atoms existed, since they could be measured with accuracy.

Third step: the atomic nucleus (10^{-15} m = 10^{-5}Å = 1 fm)

The size of the nucleus is close to 1 femtometer (fm), that is 10^{-15} m. In 1907, the English physicist Ernest Rutherford arrived at the conclusion that atomic nuclei were extraordinarily small. First, he discovered that alpha particles, emitted at high velocity by the radioactive disintegration of radium, were none other than helium atoms stripped of their electron cloud (and thus positively charged). These helium nuclei could go through thin leaves of gold as if this metal were completely transparent. However, there were exceptions: a very small number of the helium nuclei rebounded back at great speed. This meant that these rare particles had struck the nucleus of a gold atom. From the ratio of the number of particles that rebounded to the number of particles that went through, Rutherford calculated that the gold nuclei were 100 000 times as small as the gold atoms (for comparison, the Sun is only 100 times as small as the radius of the orbit of the Earth).

The complementarity of waves and particles mentioned earlier manifests itself in a strange way: it is impossible to measure with precision the speed and the position of a particle at the same time, because the product of the two uncertainties remains constant. If the speed is known with accuracy, the position is not, and vice versa. Discovered in 1926 by the German physicist Werner Heisenberg, this *uncertainty principle* has become more and more fundamental to our understanding of the strange subatomic world of quanta. It was because Rutherford had used sufficiently fast alpha particles (without knowing at the time that great speed was necessary) that he had been able to probe the size of the nucleus of the gold atom. The great uncertainty associated with the very great speed of the alpha particles produced a very low uncertainty for the size of the gold nucleus. Likewise, through the use of faster and faster particle accelerators, knowledge of nuclear structure has made great progress recently.

It soon became clear that the nucleus of the hydrogen atom (which had been dubbed the '*proton*') had to exist in several copies in heavier atomic nuclei. This was needed to balance electrically the negative charge of the larger number of electrons. However, there remained a difficulty: there were atoms with identical chemical properties, but with a different atomic mass (in 1913 the chemist Frederick Soddy proposed the name '*isotope*'). The difficulty was resolved in 1932 by the English physicist James Chadwick, when he discovered the existence of another particle, about as heavy as the proton, but electrically neutral: the *neutron*.

The order of the elements in the Periodic Table implied a regularly increasing number of protons in the nucleus (the *atomic number*) which is characteristic of its chemical properties, while the mass of the nucleus (the *atomic mass*) pointed to the addition of a varying number of neutrons. In a heavy nucleus, this number is generally higher than the number of protons, whereas in the lighter nuclei the proportion of neutrons is less. For instance, hydrogen has a single proton and no neutron. There is, however, a heavier isotope of ordinary hydrogen, called deuterium, with 1 proton and 1 neutron. If we consider heavier atoms, in oxygen there are 8 protons and 8 neutrons, but for gold there are 118 neutrons and only 79 protons. The atomic mass of hydrogen is thus 1, that of deuterium 2, that of oxygen $8+8 = 16$ and that of gold $118+79 = 197$. Since protons are positively charged, they repel each other. In order to hold them packed in the atomic nucleus, a force of a different kind is needed to overcome the electric repulsion. This force cannot be gravitation, because its attraction, at the same distance, is 10^{37} times weaker than the electric repulsion. Thus, there must be a much stronger nuclear force that falls practically to zero outside the nucleus (1 fm). It was called the '*strong nuclear force*', to distinguish it from the '*weak nuclear force*' that was later put forward to explain beta radioactivity (see Appendix A).

Fourth step: the elementary particles (10^{-18} m = 0.001 fm)

In the preceding step, matter seemed to be made uniquely of protons, neutrons and electrons. In our fourth step toward the extremely small, the electron remains an elementary particle, but the interior of protons and neutrons shows a fine structure. This fine structure was revealed, at the end of the twentieth century, by very high speed collisions achieved in large particle-accelerators. The technique is fundamentally the same as that Rutherford used in discovering the size of the atomic nucleus, but it uses energies considerably larger than for the alpha particle experiment, allowing a much finer spatial resolution.

This spatial resolution showed that the neutron and the proton each consists of three particles 1000 times smaller, the '*quarks*'. Ordinary matter (meaning that which exists around us and nearly everywhere in the Universe) is made up of only two varieties of quarks named 'up' (symbol u) of electric charge $+2/3$, and 'down' (symbol d) of electric charge $-1/3$. The proton consists of 2 up and 1 down, thus its total charge is $4/3 - 1/3 = +1$, while the neutron contains 1 up and 2 down; in this way, its total charge is $2/3 - 2/3 = 0$. It is obvious that the fractional charge of quarks arises from our arbitrary choice of unit charge $+1$ for the proton and -1 for the electron.

The level of our fourth step corresponds to the maximum spatial resolution achieved at the end of the twentieth century in very large particle-accelerators. This is why we are not yet certain of the actual size of the quarks or of the electron, except that it cannot exceed 0.001 fm. In particular, the actual dimension of the electron should not be confused with the 'cloud' around the nucleus, which arises from the uncertainty of the electron's location at very low energies.

Last steps: the plunge to 10^{-35} m

At the end of the twentieth century, our descent into the last steps is based essentially on what we can surmise from the theory of the symmetries of interactions between particles (see Appendix B).

By the mid century, *four interactions* of different symmetries had been identified, which would seem sufficient to explain *all* the forces present in nature. Two of these interactions, gravity and electromagnetism, act over very large distances, whereas the other two, the strong nuclear force and the weak nuclear force, are confined to the interior of the atomic nucleus.

Between 1960 and 1970, the physicists Sheldon Glashow, Abdus Salam and Steven Weinberg proposed a means of unifying electromagnetism with the weak nuclear force. There is only one particle which carries the electromagnetic force, namely the photon (particle of light). Because of the more complex symmetries present in the weak force, the role of the photon has to be played by *three* new particles, until then unknown: two electrically charged particles, identical except for their opposite charges, called W^+ and W^-, as well as a different, neutral particle, called Z^0. Contrary to the photon, which is massless, these three new particles need to have mass in order to confine the weak force to within the diameter of the nucleus. Since the strength of this confinement is known, it is possible to predict the mass of the particles.

This new theory explained so well all radioactive phenomena of the beta decay that the Nobel Committee had no hesitation in awarding the Nobel Prize for 1979 to Glashow, Salam and Weinberg *before* the discovery of the predicted particles, so evident was it that they had discovered the only possible correct theory to explain the weak nuclear force. Four years later, in 1983, the W^+ and W^- were observed in the large particle-accelerator at CERN (the European Council for Nuclear

Research) in Geneva with about the mass predicted (85 times the mass of the proton). This was rapidly followed by the discovery of the Z^0 particle, with a mass close to 97 times that of the proton.

These discoveries brought great respectability to the method that uses all the symmetries revealed by the interactions. At the same time it raised expectations that one could go down still deeper into the atomic nucleus, in order to achieve a 'grand unified theory'. One of the reasons for this is that the Z^0 particle would be identical to the photon if it had no mass. To explain its mass, it was necessary to merge the weak force with electromagnetism into what was dubbed the 'electro-weak theory'. This theory shows that the force constants of electro-magnetism and of the weak nuclear force, which were so different at weak energies, converge to a single value at high energies, which corre-spond to those short distances within the atomic nucleus.

There are good clues that the strong nuclear force and gravitation also converge to this unique value of the force constants, although at still higher energies. This argues for the use of the same method to unify all the forces in order to achieve the merging of all the symme-tries discovered in particle interactions. At the close of this century, all physicists hope that a theory which will unify *all* the forces of nature in a grand '*supersymmetry*' is not too far off.

Even if the theory is not yet complete, the convergence of the forces that predicts a complete symmetry at high energies suggests the cause of the initial 'flicker' that unleashed the Big Bang. It would be a quantum fluctuation inducing the spontaneous rupture of the pre-existing grand symmetry of nothingness (see Appendix F).

THE RACE TOWARD
COMPLEXITY

I would like to know how God created the world. I am not
interested in this or that phenomenon, in the spectrum of
this or that element. I would like to know His thoughts, the
rest is detail.

Einstein (quoted by Freeman Dyson, 1988)

Man must cling to the belief that the incomprehensible is
comprehensible. Otherwise he would give up investigating.

Goethe, *Maxims and Reflections*, 1829

The grand beginning

Let us summarize the initial events in their chronological order. First,
we can imagine a quantum fluctuation in the void, which began everything. The perfect symmetry of the little bubble of pure energy is unstable and breaks up spontaneously. We follow it at the instant when it is
still smaller than a proton. It inflates exponentially while creating its
space–time dimensions and, after 10^{-32} second, it is already larger than
the present Solar System. This exponential '*inflation*' (see Appendix F)
creates all the matter and all the radiation still present in the Universe.

After that, the phase transition ends. The change of state has forever
broken the initial symmetry. From now on, only the nuclear forces will
remain confined between the quarks, whereas the forces of gravity and
electromagnetism now act at a distance. The Universe will continue its
expansion in an almost linear manner, restrained merely by gravitation, right up to the present.

But where did all this energy come from, with its ability to create the enormous bulk of matter and radiation that we see in the stars and the galaxies? It was drawn from the gravitational field created by the appearance of matter. Every gravitational field is in effect a field of *potential* energy (that is, of *negative* energy) that compensates exactly for the positive energy mc^2 of the mass m that created it (c being the symbol for the speed of light). The total energy of the Universe has thus always remained *nil*.

However, if the energy of the Universe has always been nil, where could the energy of the initial little bubble come from, tiny as it was? As a matter of fact, conservation of energy is only true on average. The Heisenberg uncertainty principle, which physicists have proved accurate in all possible conditions so far measured, guarantees that there can be energy fluctuations, in particular in *empty space* (see Glossary).

How can pure energy make matter? It is an experiment that physicists can easily make in large particle-accelerators. With enough energy, a pair of particles of matter and antimatter is created, such as a quark–antiquark pair. The energy of the Big Bang was in the form of gamma-rays; while cooling down toward 10^{13} K (kelvins), these gamma-rays continued to make more matter–antimatter particles than matter–antimatter collisions made gamma-rays. Hence, during inflation, the amount of matter and antimatter never stopped growing.

Needless to say, the 'dream' of a quantum fluctuation will remain strongly speculative until a final unified theory of quantum gravity is firmly established. Only then shall we be able to understand the reason behind the symmetry breakdown that drove the inflation. However, there hardly seems to be any alternative. The known properties of the quanta involved at the instant of the Big Bang eliminate the need to consider an infinitely small mathematical *singularity* which has no physical meaning, while the Heisenberg principle explains the transient appearance of energy in the initial 'bubble'.

On the other hand, after the first 1/1000th of a second, we are on much safer ground, because the temperature had dropped low enough for particles to obey the known laws of physics. Then the quarks began to condense into droplets of mesons (pairs of quarks–antiquarks) and of neutrons, protons, antineutrons and antiprotons (collections of three quarks or antiquarks). During the cooling, large amounts of matter and antimatter were created in equal proportions. Thermodynamic equilibrium was finally established when the cooling reached 10^{12} K. At lower temperatures, annihilations took over: particles of matter colliding with antimatter were transformed into high-energy radiation. After all possible annihilations, the antimatter disappeared completely from the Universe. Yet, there remained some matter; how is that possible?

By comparing the photons observed in the *fossil radiation* (see Glossary) with the matter particles still present in the Universe, we can calculate that there are now 3 billion photons per particle of matter. Since each photon we see in the afterglow of the Big Bang was produced by the annihilation of a pair of matter and antimatter particles, it means that there must have been one extra particle of matter too many during the annihilation of 3 billion pairs of particles. The asymmetry between matter and antimatter is thus remarkably small, of the order of 1/3 000 000 000. It looks negligible, but this extra amount of matter explains our existence and that of the familiar world around us.

Where does this excess matter come from? An asymmetry in the behavior of the weak nuclear force has been discovered in the decay processes of the *neutral meson* K^0. This meson is one of the 'droplets' of quark–antiquark described before, and the asymmetry in its decay products makes a little more matter than antimatter, namely one single extra particle of matter for each 500 K^0 mesons. This asymmetry seems to be adequate to explain the excess of matter in the complex mixture of 'droplets' containing two and three quarks.

So, if the present Universe contains matter only and not antimatter, it must stem from the symmetry breakdown of the weak nuclear force. This decay asymmetry is irreversible, which means that it *initiates the flow of time* in the world of elementary particles. We can deduce an important consequence: before the first breakdown in the symmetry of forces, time did not flow away. In other words, the phrase '*before the Big Bang*' has no meaning because the flow of time did not exist. Around 1967, the great Soviet physicist Andrei Sakharov (1921–1989) discovered this connection between the matter–antimatter asymmetry and the asymmetry resulting from the flow of time.

The first fossil trace

From the first 1/100th of a second, up to 3 or 4 minutes after the Big Bang, the first very light atomic nuclei formed left a direct fossil trace.

This is shown by their relative cosmic abundances which have hardly changed. The origin of these abundance ratios would still be a mystery without the Big Bang theory.

In particular, the ratio of cosmic abundances, expressed in mass percentages, is about 28% helium to 71% hydrogen (the total for all missing elements amounts approximately to 1%). It is also known that the stars use hydrogen as nuclear fuel and transform it into helium. However, throughout the life of the Universe, the stars could not have burned more than about 4% of hydrogen into helium. If that 4% is set aside, 24% of the 'primordial' helium still remains completely unexplained unless it was formed in the second minute after the Big Bang.

Let us reexamine in detail what happened after the first few seconds. The large annihilations of matter with antimatter stopped, because the temperature had fallen below 10 billion degrees. The proton became stable; but the neutron, while remaining uncombined, continued to decay slowly into a proton, by ejecting an electron and a neutrino. The half-life of this transmutation is well known since it can be measured in the laboratory.

From the known rate of expansion, we can calculate that, a minute and a half later, the temperature had dropped to 1 billion degrees. Owing to the neutron decay, the proportion of neutrons to protons had fallen from 50/50 at 100 billion degrees, to 24/76 at 10 billion degrees, and to 12/88 at 1 billion degrees. It then stabilized because at that temperature the protons combined with all available neutrons to make deuterium (the heavy isotope of hydrogen). In turn, two deuterium nuclei transformed into helium. The theory not only explains the right proportion of helium-4 (^{4}He, isotope of mass 4) but also those of ^{3}He and ^{7}Li (measured to have an abundance 0.001% for ^{3}He and 0.000 000 01% for ^{7}Li), as well as 0.002% of residual deuterium. Figure 2.1 shows that the theoretical abundances of ^{4}He and ^{3}He do not change much, as a function of the average density of the Universe. In contrast, that of deuterium drops rapidly with an increasing average density, so that the abundance of deuterium observed in interstellar space is a sensitive indicator of the average density of the Universe. At that time, if we neglect the minor traces of some light elements, there were only two major elements in the Universe: helium (24%) and hydrogen (76%).

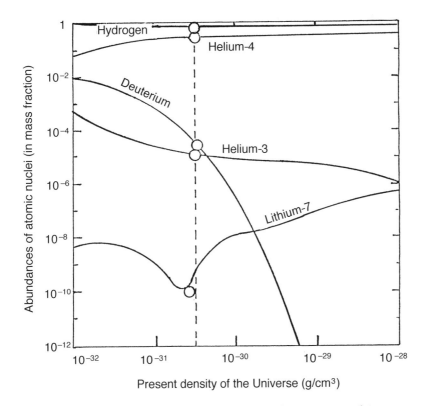

FIGURE 2.1 The relative abundances of the light atomic nuclei as a function of the present average density of the Universe, produced by the quenching of nuclear syntheses near 1 billion degrees. The five continuous curves show the computed abundances at equilibrium for different possible densities. The five circles are the actual measurements made by astronomers. The vertical dashed line presents the best possible fit of the observed abundances to the theoretical curves. It yields a result of $3 \times 10^{-31} g/cm^3$, not very different from the density of the detectable mass, but much less than the 'critical' density (see Glossary) of $10^{-29} g/cm^3$.

The Universe becomes transparent

After the abundances of the light elements were set up by the fast cooling of the primordial 'fireball', the mixture continued to expand for 300 000 years without any change of state. It remained a *plasma*; this means that the light elements were in the form of atomic nuclei mixed with free electrons. As the cooling continued, some electrons began to combine with atomic nuclei, forming *ions*; at about 4000 K, the transformation was nearly complete. Most of the ions became electrically neutral and the plasma became a gas.

In the plasma, light was scattered by free electrons; hence the plasma was opaque (exactly like the interior of the Sun). In contrast, as soon as the electrons combined with ions to form neutral atoms, the gas became transparent. At that moment, light escaped from the 'fireball' and the cosmic afterglow began its long journey to the present.

Earlier, matter and radiation were always at the same temperature, because the collisions between photons and electrons induced an equal sharing of the heat (in this situation, it is said that matter and radiation are *coupled*). Later, there was *decoupling* of matter from radiation, because the gas had become transparent. Matter now felt the influence of an extremely weak force, gravitation, while radiation was hardly affected.

But in a gas that was originally uniform, gravitation could not create inhomogeneities spontaneously; it could only amplify the preexisting ones. In order to form galaxies and clusters of galaxies, gravity had to trigger the collapse to the center of mass of an already-present clump. Were these clumps present after 300 000 years? Until 1988, the cosmic afterglow had been observed only from the ground with radiotelescopes; this glow appeared to be remarkably uniform, but atmospheric absorption prevented it from being studied in detail.

NASA set out to investigate whether clumps were already visible in the cosmic afterglow, by using a satellite in orbit outside the Earth's atmosphere. The so-called COBE satellite was built for that purpose; it observed the cosmic afterglow for two years (between 1992 and 1994). These observations verified the remarkable isotropy of the cosmic afterglow looking along several million different directions in the sky, and at more than 70 wavelengths in each direction. The distribution of intensities fitted that of a perfect radiator everywhere (what physicists call a 'black body') at a uniform temperature of 2.736 K, determined to two more significant figures than previous measurements. The sky background yielded a thermal curve so uniform and precise that another significant figure could easily be obtained if the calibrations made on the ground could be improved (Figure 2.2).

An initial correction was needed to separate the signals coming from

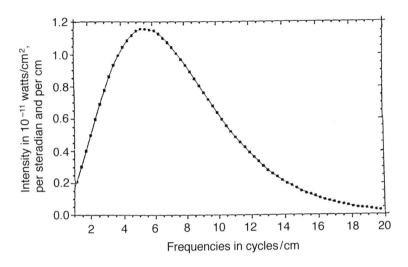

FIGURE 2.2   Frequency distribution of the cosmic afterglow observed by the NASA COBE satellite in April 1992 for the point in the sky at the north pole of the Galaxy. A million different points in the sky have all shown the same identical result, revealing the perfect isotropy of the cosmic afterglow. Each point represents one measurement, while the continuous curve represents the theoretical spectrum of a perfect blackbody radiator at 2.736 K (that is, at less than 3 degrees from absolute zero). The match of the data to the theoretical curve is remarkable. The frequencies are expressed by the number of waves per centimeter, so that the corresponding wavelength is 1 cm at the extreme left; from left to right it goes down to 0.05 mm. The maximum intensity occurs at a wavelength of about 2 mm.

deep space, from the thermal emission of the dust located in the plane of the Galaxy. Fortunately, this thermal emission is easily identified. In the first place, it is found only along the galactic equator. Secondly, it has a characteristic spectrum; this makes the correction straightforward. The residual signal, after this first correction, gave the impression that a whole hemisphere of the sky was slightly warmer, while the other hemisphere was slightly colder. This feature completely disappeared when the Doppler effect caused by the movement of the Earth relative to the distant galaxies was taken into account.

After the two corrections are made, the entire sky is then, for all

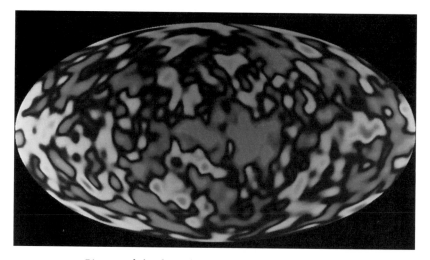

FIGURE 2.3   Picture of the deep sky at millimeter wavelengths, revealed in
1992 by NASA's COBE satellite. The whole sphere of the sky is projected
onto an ellipse, which distorts the surface area but preserves the
proportions. We see here what the Universe was like in the distant past,
before the formation of stars and galaxies. The limits of the white areas as
well as of the gray areas are, respectively, 27/1 000 000 of a degree warmer,
and 27/1 000 000 of a degree colder, than the average temperature of the
rest of the sky, represented by the black intervening areas.

wavelengths used and in all directions, at exactly the same temperature, 2.736 K. The accuracy of the *relative* measurements is such,
however, that immense 'clouds' of varying sizes appeared, all about
27/1 000 000 of a degree warmer than the average temperature, while
the gaps between the clouds were about 27/1 000 000 of a degree colder
(see Figure 2.3). Although extraordinarily slight, these differences are
real because they are beyond instrumental noise at a statistical level.

The angular resolution of the instrument was only 7° of arc.
However, analysis of the angular distribution of the temperature
fluctuations, as well as the steady value of their weak amplitude, show
that the observations are consistent with the inflation theory. As a
matter of fact, the size distribution of the 'clumps' remains the same at
all magnifications. This is a 'scale invariance' characteristic of *fractals*.

Such scale invariance is predicted for the quantum fluctuations that could have started inflation. These fluctuations would have been increased to the size of clusters of galaxies by the immensity of the expansion resulting from inflation. Not only does the size distribution of the clumps remain the same at all magnifications, but also the density distribution, inferred from the observed temperatures, satisfies the models proposed for the formation of the clusters and superclusters of galaxies.

The amplitude of the density fluctuations cannot yet be explained, because a grand unified theory of elementary particles has not been developed yet. The theory based on a minimal *SU(5) symmetry* (see Appendix B) yields results that are much too large. But for other reasons we know that this symmetry group is not satisfactory; in particular, it predicts a proton lifetime contradicted by other experiments. Hence the discrepancy arises from the difficulty of not having a grand unified theory, and not from the COBE results, which show how the superclusters of galaxies got their sizes out of the 'clumps' observed in the cosmic afterglow.

In April 1992, at the time of the official announcement of the first COBE results, the media stated that these observations had brought the final proof of the Big Bang. This statement was superficial and basically wrong. The arguments for the Big Bang were not new, and no observations had really changed the situation. What COBE displayed for us was a picture of the deep sky background, showing what the Universe looked like in the distant past, before the formation of stars and galaxies. The big news was that, 300 000 years after the Big Bang, a scheme for the future structure of the Universe was already faintly visible. The COBE satellite detected clumps of gigantic size, already slightly differentiated in density, that were ultimately the seeds of clusters of galaxies.

These immense irregularities reassured astronomers, who were wondering how clusters of galaxies formed if there were not already some preexisting clumps. Moreover, at the same stroke, their scale invariance explained their origin: they were extremely small quantum fluctuations extraordinarily amplified by inflation. Needless to say this attractive possibility remains to be confirmed by an adequate theory of quantum gravity.

The essential point is that we have observed the intensity and the scale of fluctuations of the density of matter as they were distributed in the Universe 300 000 years after the Big Bang, and that we can now attempt to deduce how the clusters of galaxies and the galaxies themselves formed.

The origin of the clusters of galaxies

The size distribution of the fluctuations observed by COBE displayed a 'scale invariance'. This is an important idea mentioned earlier, but worth clarifying. The distribution of irregularities looks alike on all scales; hence the general appearance of the details remains the same for any fraction of the picture. This corresponds to the definition of a 'fractal', a neologism introduced by Benoit Mandelbrot in 1975.

The distribution of galaxies is a fractal. If the trend of the galaxies is to associate into clusters, the clusters of galaxies themselves try to associate into superclusters, and the progression stops only because it is limited by the size of the visible Universe (see Figure 2.4).

However, the clumping detected by COBE does not seem to produce density differences that are large enough to account for the formation of galaxies in a reasonable time. The characteristic time for the gravitational settling near the center of a clump is easily computed. For a fluctuation of 1.7%, it is of the order of 1 billion years. That would be satisfactory. However, if the fluctuations are 0.3%, the settling requires 13 billion years; for 0.2%, the process would take more time than the present age of the Universe. And here is COBE showing us temperature fluctuations of only 0.001% – 1000 times too small!

Fortunately, there is a way out. The actual density of the Universe could be many times larger than our assumptions, making for a much faster settling. This leads us on to the problem of the 'missing mass' mentioned earlier. The bulk of the missing mass cannot be ordinary matter (that is, matter made up of protons, neutrons and electrons), otherwise the curves of Figure 2.1 would be changed and the convincing coincidences shown by the dashed vertical line would disappear. But there are several other possibilities. Leaving aside the more exotic particles proposed by theorists, but not yet discovered, there are the three types of neutrino. The mass of these neutrinos was believed to be

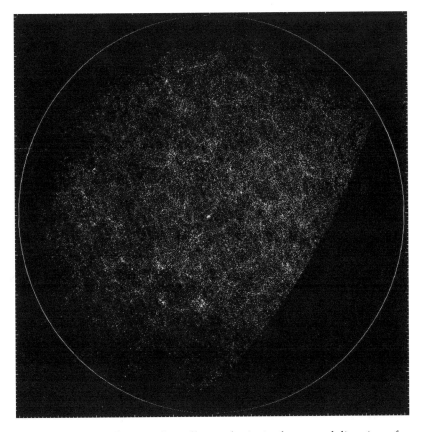

FIGURE 2.4   Distribution of 1 million galaxies in the general direction of
the north pole of our Galaxy. Each point represents a galaxy density
(needless to say all nearby stars have been removed from the picture).
Toward the center, a cluster of galaxies, which stands out as a white
patch, is the well-known cluster of galaxies in the Coma constellation.
The large cluster below is the cluster of galaxies in Virgo. Regular in
general, the details reveal a spatial distribution of structures with
fluctuations and irregularities.

nil until recently, because we could not measure it. However, there are
recent measurements that suggest a very small mass for at least one
type of neutrino. Invisible neutrinos are so abundant in the Universe
that the missing mass could be explained if one of the neutrinos had a

mass 30 000 times lighter than that of the electron (which is itself 2000 times lighter than the proton and the neutron). Such a huge amount of neutrinos was produced in the Universe by the huge number of matter–antimatter annihilations, which is itself predicted by the huge amount of radiation measured in the cosmic afterglow. Recent laboratory estimates are consistent with this mass of the neutrino. This density of matter in the Universe would enable clusters of galaxies to form in 1 billion years, a time span compatible with the age of the Universe.

The reason for this rapid formation of the galactic clusters is that the neutrinos decouple from matter a long time before the photons do. They would have formed early invisible clumps, acting as centers of mass. These centers of mass would have attracted hydrogen and helium early in the process, as soon as they were decoupled from radiation 300 000 years after the Big Bang. Some other problems remain, but solutions can be glimpsed.

In the model that we have used, quantum fluctuations were amplified to an extreme by the decrease of space curvature during inflation. This is why radiation and matter show the same perturbations (they are intrinsic to space curvature). However, before the decoupling of matter and radiation, in the ionized fireball, interactions between photons and electrons wiped out those fluctuations of mass that were too small (an effect of what the specialists call the 'viscosity' of the photons). Computation shows that the remaining perturbations correspond to a mass of at least 10^{13} times the mass of our Sun. The minimal mass involved is therefore of the order of a thousand average galaxies, which corresponds well to the mass of a cluster of galaxies.

The densest spot of an enormous mass of this type easily exceeds the *critical density* of the Universe. This means that such a mass is not going to participate indefinitely in the expansion of the Universe. It is going to slow down to a maximum size, then fall back upon itself. The collapse is more rapid along the smallest diameter: hence, there is a tendency for flattened objects to form. The Soviet cosmologist Ya. B. Zel'dovich has called these flattened objects 'crêpes' (Figure 2.5). More elongated objects can also form in the shape of 'cigars'. Soon cooled down by the radiation process, these objects form clusters or super-clusters of galaxies.

I have just described how the inflation theory, combined with the

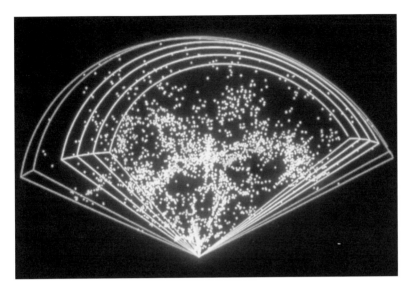

FIGURE 2.5   Distribution of the galaxies in depth, as measured from our
Galaxy, which is located at the center of the diagram. The figure is based
on a study carried out at Harvard. The distribution shows voids,
filaments and two-dimensional structures that can be associated with
the so-called 'crêpes' of Zel'dovich.

mass of the neutrino, allows us to avoid non-orthodox ideas that
invoke topological faults in the fabric of space–time (like 'walls' or
'cosmic strings') used by theorists to 'seed' the locations where galaxies will form. The *absence* of these hypothetical anomalies in
space–time (to which we should add the 'magnetic monopoles'),
assuming they exist, could be explained by inflation. Inflation dilutes
all these anomalies to an unimaginable degree, blowing away monopoles, strings and cosmic walls outside of the domain which has
become the observable Universe.

After the first 1 or 2 billion years, we reach the epoch when the first
superclusters of galaxies were already well separated. The fractal
nature of the density fluctuations means that the superclusters split
into clusters, and the clusters into galaxies. Only at the scale of single
galaxies were the fluctuations damped in the fireball, before it became
transparent. Hence the hierarchy of superclusters and clusters stops at
the dimension of small galaxies.

Let us describe the Universe as it was at that time. Its very large structure did not change much any more, apart from expansion which enlarged its size by a factor of 10 and its volume by a factor of 1000. The small blobs coming from quantum fluctuations were now steadily enhanced by the force of gravitation. Observations show that two-dimensional structures (resembling Zel'dovich's 'crêpes', see Figures 2.4 and 2.5) often seem to be distributed over the walls of 'bubbles' whose center is empty. In particular, a thorough investigation of 20 000 galaxies was conducted by Margaret Geller, John Huchra and their colleagues at the Smithsonian Astrophysical Observatory at Harvard. A small part of their results is shown in Figure 2.5. This work raises a serious issue for theorists trying to understand why the 'crêpes' are arranged over the walls of 'bubbles'. But for now, we are not attempting to explain everything; we strive only to describe the global evolution of the Universe in a continuous and coherent fashion.

After the first billion years, the galaxies were still only huge masses of hydrogen and helium that whirled and cooled. The supercluster structures had masses of the order of 10 000 galaxies like ours. Within the superclusters, galaxies agglomerated in clusters ranging from 30 to 1000 galaxies.

The size of the superclusters, then of the clusters of galaxies, and finally that of the galaxies themselves, was set by the way the spacings between the clumps of the primordial fireball opened up and became transparent. Later on, the outer regions of the galaxies themselves cooled off to less than 4000 K; no longer ionized, these regions became transparent, allowing the inner heat of the galaxies to radiate into outer space. But a cooler galaxy implies a falling gas pressure, and the weight that this pressure could withstand diminished. Eventually, blobs of gas of diminishing masses can collapse, down to 10^7 and even to 10^5 solar masses. In each galaxy, hundreds of collapsing gaseous nodules appear in this size range, announcing the formation of globular clusters of stars.

The birth of stars

Let us follow what probably occurred in a galaxy like ours. In a large, barely flattened ball of gas, whose diameter was several hundred thousand light-years, hundreds of smaller gaseous globules formed and slowly concentrated. The largest of these gaseous clusters appeared at the center of the Galaxy. It is likely that several smaller globules, presumably assisted by the central force of gravitation, collided and aggregated to form the central core, while a couple of hundred smaller gaseous globules remained spherically distributed. At that time, the cosmic afterglow, diluted by the expansion of the Universe, had already cooled to less than 40 K, hence the galactic gas could continue to cool by radiating further and further into the infrared. In all the globules, gravity was soon counterbalanced by residual gas pressure, but this pressure diminished with falling temperature, allowing smaller and smaller masses to collapse gravitationally.

Within the central galactic core, as well as in the gaseous globules, hundreds of millions of much smaller globules appeared. Their sizes are again described by a fractal distribution, limited toward smaller sizes by gas cooling, which could not go beyond the general cooling of the cosmic afterglow, set by the expansion of the Universe.

In each gas nodule, there was a time when its central density was large enough to make it collapse upon itself. The inside layers collapsed first, the outer layers later. While falling faster and faster, the gas was eventually stopped at the center by the masses of gas that arrived from all directions. This transformed the kinetic energy of the infall into heat. The larger pressure of the gas, developed by heat, became able to support the weight of new layers of gas, and soon the central gaseous sphere was so hot that it radiated a reddish glow.

We have just described the birth of a star.

However, the turbulent gas motions prevented a large part of the available gas from accumulating directly onto the newborn star. The angular momentum of this gas caused it to whirl and form large rings that orbited around the new star. In this way, an *accretion disk* was formed. This is a very general phenomenon, essential for any star to grow and reach its final mass. The rings were stable only if their rotation conformed to Kepler's laws. This means that the outer parts of the disk turned more slowly than the inner parts. Because of this differential rotation, there was friction between the different zones of the disk, arising from the viscosity of the turbulent gas. This slowed the interior rings, which were driven to spiral toward the central star and eventually captured by it. In return, the outer rings accelerated, increased their angular momentum and spiralled out to greater and greater distances.

The fraction of the nodule that was late in its fall progressively thickened the rings. Part of it joined the inner rings and eventually added mass to the central star, while the remainder was ejected toward the outer rings. This mechanism explains how a gas nodule, whose size is several light-years, can concentrate a good portion of its mass into an object as small as a star, of diameter ten million times smaller, without making it spin too fast.

Sometimes the collapsing nodule was too massive, for instance from 50 to 100 solar masses. Sometimes also, due to turbulence, the accretion disk spun too quickly. Then it took on an asymmetric form. A density wave traced out an elongated S that eventually turned into a bar and ended by fragmenting into several stars. This mechanism explains on the one hand the absence of stars with mass much larger than 100 solar masses, and on the other hand the presence of a large number of double or multiple stars, gravitationally bound to each other.

The mechanism of the accretion disk is a general process of star formation that was slow to be understood, and was only generally accepted after the observations by NASA's IRAS infrared satellite revealed the ubiquity of accretion disk remnants around a very large number of extremely young stars. The evolution of accretion disks has also turned out to be the origin of the planetary systems. Even before the recent discovery of other planets around nearby stars, accretion disks supported the belief that there must be innumerable undetected planets around stars.

What is the origin of the mass distribution among stars? We have already mentioned that this distribution is a fractal. It is produced by the scale invariance in the turbulent motions of the gas forming the nodules. The available mass distributes itself equally through all possible scale sizes. Therefore there will be countless small stars whereas the more massive stars will be much less numerous.

In principle, the scale invariance in the mass distribution would have to lead to 10 times more stars that are 10 times less massive, 100 times more stars that are 100 times less massive, etc., with important statistical fluctuations of course. Such a distribution is in approximate agreement with star counts in the vicinity of the Sun.

After a couple of billion years in the evolution of our Galaxy, there were already hundreds of millions of stars in the huge cluster at the central core as well as in more than 100 peripheral globular clusters. These globular clusters revolved around the central core on elongated elliptical orbits distributed in a huge spherical halo. In each of these globular clusters, millions of stars were held tightly together by gravitation. Their quasi-perfect spherical symmetry has only improved through the ages, due to the averaging effect of gravitational interaction among all stars. At the present time, globular clusters still contain by far the oldest stars in our Galaxy (see Figure 2.6 and Appendix D).

After formation of the globular clusters, the largest portion of the still available mass nevertheless remained in the diluted gas separating the clusters. This mass of gas was initially distributed in a huge quasi-spherical volume. This volume diminished in size and flattened out as it revolved faster and faster and as it concentrated to a smaller volume. As it speeded up, the rotation produced density waves that spiralled in around the center. These spiral waves whirled faster and faster as they came nearer the center. The largest part of the Galaxy ended up forming a very flat disk, whose spiral density waves eventually began to form stars, repeating billions of times the process described above.

The stars formed in the galactic disk are younger than those formed either in the central cluster or in the globular clusters. Within the clusters, no new stars form any more because all the gas available has been used up in making the early stars. In contrast, star formation continues for a very long time in the spiralling waves of the disk, which propagate while rotating through the mass of gas still available.

Eventually, globular clusters became isolated systems bound by gravity on elliptical orbits around the center of the Galaxy; they alternately dive toward the galactic center, then stretch out into the halo, with periods of revolution of hundreds of millions of years.

The radius of a cluster is very well defined, probably because at each pass near the galactic nucleus, the clusters experience maximal tidal forces. The stars at the edges are set loose. The remaining stars are held within a very compact volume. In the past, near passes were not rare within a cluster; they would often eject one or two stars along hyperbolic orbits. This effect, as well as the galactic tidal forces, explain why we can observe the presence, in the galactic halo, of a certain number of isolated stars that are as old as those in the globular clusters. Within the clusters, the distribution of speeds shows that now all stars have

FIGURE 2.6   A globular cluster of stars: the globular cluster in Hercules. To the naked eye, this cluster appears to be a single star of fifth magnitude. It is one of the 125 globular clusters that are present in the spherical halo surrounding our Galaxy. These globular clusters contain the oldest stars of our Galaxy (see Appendix D).

reached a dynamic steady state; the lightest stars travel faster and venture farther from the center than the heaviest ones.

Finally, in 1992 observations with the Hubble Space Telescope showed the presence of some blue, very young stars (see Chapter 3) in the deepest hearts of some globular clusters. The insufficient resolving power of earlier ground-based telescopes had prevented this discovery. The only possible interpretation is that the birth of these blue stars was triggered recently by the collision of two old stars whose merger allowed the restart of the nuclear combustion of hydrogen into helium.

The central galactic cluster is not essentially different from a globular cluster, except for its larger size and its more numerous stars. The greater numerical density of these stars implies a large collision rate. In the long run, these collisions must have accumulated enough mass on a single object to make it disappear into a gigantic *black hole* (see next section) early in the history of the Universe. Such a phenomenon must have occurred in a large number of galaxies. The presence of a giant black hole at the center of our Galaxy, although difficult to prove, has become very probable because several new clues found in 1996 point to it.

The presence of giant black holes at the center of a multitude of galaxies is an almost inevitable consequence of the formation of a large central cluster of stars. Paradoxically, these black holes became at a very early stage the brightest objects ever observed in the Universe. But before discussing the evolution and fate of the stars, we should review what is known about these astonishingly bright objects: the quasars.

The quasars

In the 2 or 3 billion years that followed the Big Bang, a spectacular phenomenon occurred in the core of a large number of galaxies: the central cluster became a beacon brighter than 1000 galaxies like ours (which now has a brightness equivalent to 100 billion suns). Suddenly, the Universe was lit by millions of these objects: the quasars. These beacons appeared and shone for some billions of years, but they seem to have almost disappeared at the present time, or at least, they do not seem to stand out any more from the nuclei of 'active' galaxies.

If we are still able to study the quasars, it is because when we look very far away, we observe the past. With large telescopes we observe more than 1 million quasars in all directions of the sky, but only in the very furthest depths of space. At first glance, they look like ordinary stars, except that their very blue light is a clue to their extremely high temperature. They were first discovered as a source of intense radio emission. The riddle of their distances was solved in 1963 by an American astronomer of Dutch origin, Maarten Schmidt. The visible spectrum of the quasars shows a general shift towards extremely long wavelengths: this shift to the red moves the ultraviolet lines of hydrogen into the *visible* region of the spectrum!

Interpreted as a Doppler shift, this *red shift* means that the quasars are receding from us at a velocity which is a substantial fraction of the speed of light. The Hubble expansion law, established for ordinary galaxies, can be used here in reverse: large red shifts show that quasars are all at astonishingly large distances, of the order of 10 billion light-years or more. To be seen at such large distances, a quasar must be brighter than billions of stars! What could be the source of such a bright light? A single explanation presented itself, and no other possibility has yet been found; to provide enough energy for such a beacon, there had to be bright rings of very hot matter in the process of feeding supermassive black holes.

A *black hole* is a region where gravitation has become so strong that anything approaching within a certain radius will no longer be able to escape, even a ray of light. The critical radius is called the 'Schwarzschild radius'. The interior of a sphere whose radius is the Schwarzschild radius is totally cut off from the rest of the Universe, and the existence of the black hole can be detected only by its gravitational attraction. Beyond the Schwarzschild radius, the gravitational attraction of a black hole behaves in every respect like that of an ordinary star of equivalent mass.

Like most stars, a black hole develops and grows from an accretion disk, but a black hole produces a gravitational 'well' much deeper than that of a star. For this reason, the portion of the disk very close to the Schwarzschild radius becomes extraordinarily bright, since it plunges deeply into that 'well' before spiralling into the black hole.

It was the German astronomer Karl Schwarzschild who, just before his death in 1916, established the theoretical existence of the black hole, as an analytical solution to the equations of general relativity, which had just been worked out by Einstein. It was and still is the only known exact analytical solution to Einstein's gravitational equations. Its form is extremely simple because of the spherical symmetry of the black hole. In particular, the Schwarzschild radius is equal to 3 km multiplied by the number of solar masses trapped inside the black hole.

The accretion disk already described for stars is a mechanism that accretes mass onto the central star, while shedding the excess angular momentum that would make the central mass spin too fast. The formation of a black hole is no different. Nearby matter is drawn by the gravitation of a central mass. At the beginning, this central mass could be that of an ordinary star. It becomes a black hole only by the accretion of a large enough amount of matter. Nearby matter goes on collapsing onto the accretion disk. This disk can be imagined as rings similar to those of Saturn, but larger and much thicker. Close to the Schwarzschild radius but still outside, the gravitational potential becomes huge. Half of this gravitational energy accelerates the ring, while the other half is converted into heat. Hence the inner rings heat up to a high temperature and become extremely bright, before spiralling and disappearing into the black hole.

Close to the Schwarzschild radius, the rings rotate at a speed 71% of the speed of light, and are heated to more than 100 million degrees, thus emitting mostly gamma-rays and X-rays. Somewhat further away, the rings are less hot, and emit mainly in the ultraviolet, which, after the shift toward the red due to their receding velocities, produces that blue light so typical of the quasars.

Before being swallowed up into the black hole, a good fraction of the ring's mass is transformed into radiation energy. The black hole grows each time it absorbs mass, yet the ring continues to broaden, which enables it to gather new matter. The high concentration of stars present in the central cluster of most galaxies makes it very likely that several black holes originally appeared and became large enough to collect individual stars in their accretion disk; they ended up by colliding with one another and merging into a gigantic central black hole collecting the mass of 10 to 100 million stars, reaching in the process a size comparable to our whole Solar System. Then, the size and brightness of its rings combined to make, for a while, one of the most luminous objects in the Universe.

When the rings do not collect enough matter to compensate for what they pour into the black hole, they thin down and their brightness diminishes. After perhaps 100 million years, the quasar ultimately vanishes for lack of matter. Of course, the central black hole lives on, but it becomes more and more difficult to detect; what is left behind is the gaseous ring (broadened and cooled) which carried away the excess angular momentum of the inner rings (Figure 2.7).

Sometimes an unexpected flux of matter, released by the collision of two galaxies for instance, begins to feed the rings again and rekindle the quasar. Since the Universe was smaller in the distant past, there was a higher rate of galactic collisions; there are fewer collisions now and the periodic rejuvenation of quasars appears to be becoming rarer and rarer.

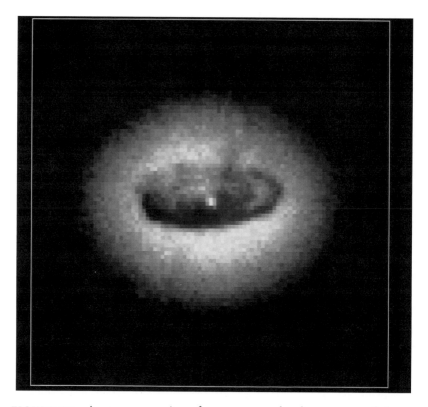

FIGURE 2.7   An enormous ring of gas seems to be the remnant of the accretion disk that fed a gigantic black hole at the core of the galaxy NGC 4261, in the cluster of galaxies of the constellation of Virgo. This picture was taken by the Hubble Space Telescope in 1992. In June 1994, the Hubble Space Telescope measured the size and rotation rate of a similar ring of gas in galaxy M87 in order to compute the mass of an invisible central object. The result established for the first time the positive existence of a supermassive black hole at the center of another galaxy.

Other objects, less bright than quasars, seem to share some of their properties. For instance, there are galaxies with an active central nucleus; there are also compact sources of radioelectric energy, sources of X-rays or of gamma-rays, that show fast fluctuations and bursts of energy. They all imply the presence of sources that appear small enough to be associated with black holes. The most likely source for all these phenomena (cosmic rays, gamma-rays, etc.), which imply the presence of high energy, is the deep potential well, just outside the Schwarzschild radius of a black hole. For instance, among the active galaxies, the 'Seyfert' galaxies emit X-rays and, in the case of NGC5548, the X-rays exhibit a periodicity of 8 minutes. The source of the X-rays would have to be a very hot object immersed in an accretion disk with an orbital period of 8 minutes around the central black hole. Such a black hole would have to have a mass close to 1 million solar masses and the temperature needed for the X-ray emission confirms the mass of the black hole.

The great distance of the quasars makes them difficult to study. We are only beginning to understand them and the future may reveal surprising new information. However, the detailed models that describe quasars as the bright rings that surround giant black holes in the cores of distant galaxies so far do not contradict current observations and indeed seem to explain all the observable properties.

It is likely that the black hole present in the core of our Galaxy had a quasar in the past, but it is unlikely that it ever reached more than 1% of the brightness of the brightest quasars. Its outer ring widened up to a distance where it stabilized, and where it can still be observed as a huge gaseous ring of molecular hydrogen. Its rotation rate was measured. From Kepler's laws, it can be deduced that it moves around an invisible mass as large as 2 million solar masses. Other clues that are in favor of the presence of a huge black hole are a point-like radio source, observed near the ring center, and the X-ray emission characteristic of the annihilation of electron–antielectron pairs. The presence of this black hole was finally accepted as having been proved at the end of 1996.

In June 1994, the first definite existence of a giant black hole at the core of another galaxy was established. The Hubble Space Telescope was able to measure both the rotation rate and the diameter of a giant gaseous ring circling the core of galaxy M87. From Kepler's Third Law, the mass of the central object is 2 *billion* solar masses; the orbit is small enough not to leave room for anything but a black hole. Since then and until very recently, more and more giant black holes have been positively proved to be at the core of several galaxies. Their presence reinforces the conviction that our understanding of the origin and nature of quasars, as described in the previous pages, is correct.

THE STELLAR ALCHEMY OF METALS

Matter is made of four elements: Earth, Water, Air and Fire.

Empedocles, c. 450 BC, quoted in Encyclopedia Britannica

In the Earth's interior, the transmutations of the 4 elements produce 2 exhalations: the dry one separates fire from stones, the moist one separates steam from metals.

Aristotle, c. 360 BC, quoted in Encyclopedia Britannica

Water is the base of metals like silver and gold, Earth, that of stones, whether precious or common; the solidification of metals is due to heat; that of stones, to cold.

Theophrastus, c. 300 BC, quoted in Encyclopedia Britannica

The life of the stars

At the time when the first quasars showed their dazzling brilliance, there were still no atoms of carbon, nitrogen, or oxygen, no metals in the Universe, nor any solid stuff (earth, rocks, etc.). It is tempting to think that the extreme temperatures reached in the accretion disks of quasars would have already produced new elements. But we must remember that the rings that reached a sufficiently high temperature were eventually absorbed by the black hole and disappeared from our Universe before having had a chance to make complex molecules.

The first stars formed out of a gaseous mixture of about 76% hydrogen and 24% helium, plus a few traces of rare light elements, such as lithium, or rare isotopes like deuterium or ^{3}He.

Table 3.1. *Major thermonuclear reactions within stars, classified by increasing ignition temperature, from 10 million degrees for hydrogen, up to 6 billion degrees for iron*

The minimum stellar mass able to reach this ignition temperature is indicated in solar mass units $M_\odot$. Only the dissociation of iron by gamma-rays is endothermic. This cooling of the stellar core induces an explosion into a supernova. The table is limited to the major reactions that involve the nine most abundant elements in the Universe. The 83 missing elements of the Periodic Table account for 0.1% of the total mass of matter. (n is the neutron.)

Thermonuclear reactions in stellar core	Resulting chemical element	Ignition temperature (million K)	Minimum mass in solar units ($M_\odot$)
Combustion of hydrogen			
$4(^1H) \rightarrow$	^4He	10	0.1
(proton–proton reaction)			
Combustion of helium			
$3(^4He) \rightarrow {}^8Be + {}^4He \rightarrow$	^{12}C	100	2
$^{12}C + {}^4He \rightarrow$	^{16}O		
Combustion of carbon			
$2(^{12}C) \rightarrow$	$^4He + {}^{20}Ne$	600	4
$^{20}Ne + {}^4He \rightarrow$	$^{23}Mg + n$		
Combustion of oxygen			
$2(^{16}O) \rightarrow$	$^4He + {}^{28}Si$	1500	8
$2(^{16}O) \rightarrow$	$^{24}Mg + 2(^4He)$		
Combustion of silicon			
$2(^{28}Si) \rightarrow$	^{56}Fe	4000	15
Photodissociation of iron			
$^{56}Fe \rightarrow$	$13(^4He) + 4n$	6000	20

This gaseous mixture split up into fragments of widely different sizes, leading to the formation of stars of varying inner temperatures, which 'ignited' many different nuclear reactions. The larger the mass of the star, the higher was its inner temperature, and the larger the atomic mass of the newly synthesized chemical element (Table 3.1).

Different stellar processes eventually scattered into space all the new possible elements, opening the door to chemical reactions. At this stage several steps along the road to complexity have been traversed, and will eventually lead the way to silicates (stony matter) and life (organic chemistry). But there is still a long way to go!

The history of a star can be summarized by envisaging a gravitational collapse toward its center, interrupted by very long stable periods, every time that the temperature, rising under gas compression, leads to the ignition of a new nuclear reaction. This begins with hydrogen, which the compression heat ignites first. Hydrogen 'burns' into helium when it reaches 10 million degrees. Deuterium burns even sooner, but there is so little of it that we can neglect its presence.

The combustion of hydrogen releases a large amount of heat which can make a star radiate for millions or billions of years, depending on its size. But the hydrogen of the outer layers does not burn, because it never becomes hot enough. This is the reason why the combustion was self-limiting and added only 4% to the 24% of primordial helium.

Despite the complexity of stellar evolution, which includes a large number of branches that led to wholly different phenomena, we have elucidated nearly all the details of the birth, life and fate of the stars. It is one of the triumphs of twentieth century astrophysics that we understand stellar evolution so well.

But our goal here is only to grasp how stellar evolution, through the synthesis of chemical elements, opened up the possibility of organic chemistry, which in turn led to the dawn of life on Earth. However, we must first justify our trust in the interpretation of stellar evolution. It is based on an extensive knowledge of the internal structure of the stars and on the impossibility of explaining the long life of the Sun (and by extension, that of the other stars) without involving thermonuclear reactions. Most of these reactions have now been studied quantitatively in the laboratory, in particular with huge particle-accelerators. These machines produce, just for an instant, collision temperatures of up to hundreds of billions of degrees, which is hotter than the cores of the hottest stars.

Moreover, we understand the gas laws well enough to know what occurs when a sphere of gas collapses under the weight of its own gravity. To make a star, this sphere must first be sufficiently massive; then compression allows it to reach a central temperature large enough to ignite nuclear reactions. Computation shows that, in order to reach the 10 million degrees needed to ignite hydrogen, a mass of at least 8%

of that of the Sun is required. It is easy to confirm that this corresponds to the size of the smallest visible stars in our neighborhood.

The brightness of a star tells us its rate of energy consumption, and thus the rate of its nuclear reactions, which in turn yields the central temperature. The radiation transferred from the center to the surface can then be computed, taking into account the opacity of the different layers of gas. The mechanism that controls the temperature of the center by means of surface radiation is a natural thermostat that operates on the same principle as the one that controls the temperature of domestic central heating systems. When the radiation drops, the star compresses, and its central temperature rises, increasing the reaction rate. This is a negative feedback which controls any tendency to modify the steady state, and explains why most stars remain stable and steady for most of their life.

In order to verify whether our deductions are correct, we can compare the mass, size and radiation rate of a large number of stars. For such comparisons, the so-called 'Hertzsprung–Russell' diagram has been very useful; it was introduced at the beginning of the twentieth century, independently in two very similar forms, by the Danish astronomer Ejnar Hertzsprung and the American astronomer Henry Norris Russell. This diagram (see Figure 3.1) charts the absolute luminosity of a star (with the solar luminosity taken as unity) as a function of its surface temperature (in kelvins).

As long as they burn hydrogen into helium, the stars fall along an oblique line of the diagram, called the 'main sequence'. The main sequence is nothing more than a classification of stars according to their mass. In Figure 3.1, the position of stars of different masses is shown in solar mass units $M_\odot$. For instance, a star of 30 solar masses is about 10 000 times brighter than our Sun, that of 10 solar masses is 1000 times brighter, and that of 3 solar masses nearly 100 times brighter. The brightest known stars are of 60 solar masses and 100 000 times brighter than our Sun, whereas the least bright stars reach less than 0.1 solar mass and 0.001 of the Sun's brightness.

One can then compute the age of each star when it has burned into helium all the hydrogen available in its high temperature core. The result depends much on the star's mass. A more massive star is more compressed by its weight; hence its core reaches a higher temperature. For a rather weak temperature elevation, the nuclear reaction rate

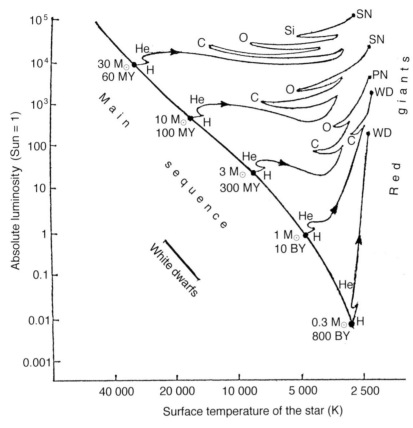

FIGURE 3.1 Theoretical Hertzsprung–Russell diagram. The lifetime on the main sequence of each typical star is in million or billion years (MY or BY). Masses are in solar mass units $M_\odot$. Arrows show the fast evolution of stars toward the red giant branch, in search of a new steady state after depletion of hydrogen in their cores. Ignition of the successive elements is shown by their chemical symbols. Return of a path toward the left indicates the compression of the star's core to a higher temperature, after the exhaustion of a given nuclear fuel. On the right, the climb in luminosity is conspicuous, mostly for low masses. It is the result of convection from the deepest layers, which provides more efficient heat transfer to the outer layers. The final states are: WD, white dwarf; PN, planetary nebula; SN, supernova. After dissipating their extended atmospheres, the naked white dwarfs reappear at the lower left of the diagram.

increases very much; hence the star's luminosity is much higher, explaining the previous results.

When a star has converted its whole core into helium, the nuclear reaction stops, and the core pressure falls. As soon as this occurs, the star's natural thermostat readjusts its temperature in search of another source of energy. The weight of the outer layers compresses the core, which collapses inward until it reaches 100 million degrees. This is the ignition temperature for helium. Then three helium nuclei combine to form one single carbon nucleus, with the release of much nuclear energy. The star leaves the main sequence; it readjusts its internal steady state while extending its envelope and reducing its surface temperature. It has now passed over into the realm of the red giants.

The branches in the path that decide the ultimate fate of the stars depend on their mass. In order to understand the nature of the branches, let us follow the evolution of a few characteristic stars. In keeping with this chronological history of the Universe, we will begin with those massive stars that evolve faster.

The fast evolution of massive stars

Nodules of 100 solar masses fragment in general into double or multiple stars. However, if a star of 100 solar masses should form, its central temperature would be so high that its radiation would blow off the outer layers into space (this is called a *stellar wind*). This wind would soon reduce the star's mass until it stabilized near 60 solar masses. Such stars of 60 solar masses are indeed the largest that are presently observed, so I will call a 'massive star' any star between 3 and 60 solar masses. Among the massive stars, I will describe in detail the behavior of a star of 30 solar masses, then more briefly that of two stars of 10 and 3 solar masses, respectively.

Star of 30 solar masses

The star's birth begins with the collapse of a nebular nodule that results from the fragmentation of a vast cloud of 10 000 to 100 000 solar masses. Each fragment subdivides into several blobs that display the pattern similarity characteristic of fractals. A small protostar forms at the center of the collapse, but the vortex that preserves angular momentum creates a large accretion disk around the proto-star. For a long time, nebular matter continues to fall on the disk, where it is stopped by a shock wave. Within a dozen million years, the disk has accreted 30 solar masses to the central star, and has broadened into a ring of gas that will last for some time.

All of this is not fundamentally different from the process that formed quasars, except that the masses and the scale involved are a good deal smaller, and the maximum temperature of the inner ring barely exceeds a few thousand degrees.

The star separates from the rings, which are cooling down because they are no longer fed from the outside. It is not impossible that these rings could eventually form totally gaseous planets; but rocky planets cannot form yet because, in the first billion years after the Big Bang, the chemical elements needed to form rocks do not yet exist. We have neither proof nor observation of any hypothetical gaseous planet which consists only of hydrogen and helium.

The central star settles and compresses its interior, which heats up. Near 50 000 degrees, the atoms of hydrogen and helium enter into collisions that are violent enough to strip them of all their electrons, thus forming a plasma. This plasma is a mixture of atomic nuclei of helium and hydrogen in a gas of free electrons. But the compression of the plasma continues. When the star center reaches 10 million degrees, the hydrogen nuclei collide with each other so violently that they no longer rebound like billiard balls: they weld into a deuterium nucleus, while liberating an electron and a neutrino. The neutrino escapes from the star, and the electron mixes with the rest of the plasma.

This reaction liberates a large amount of energy in the form of gamma-rays. These photons are going to diffuse slowly while heating the star; they will end by escaping in the form of visible light. The deuterium that has just been formed encounters another proton and forms helium-3.

Later, the collision of two ^{3}He atoms frees two protons and creates a ^{4}He nucleus. This chain of reactions has variations; it can go through transient nuclei of lithium (Li) and beryllium (Be), but the final result remains the same. Everything is as if four protons combined into a single ^{4}He nucleus, with the release of much energy; this progression is called the proton–proton reaction. Much later, when a new generation of stars can use heavier elements, other reactions will change the production rate of energy; but at this point, the other elements do not yet exist in the Universe.

When the nuclear reactions begin at 10 million degrees, the collapse goes on for a while and the star continues to settle. It is only when the star's central temperature reaches 25 million degrees that the production rate of energy becomes large enough for its radiation to support the weight of the outer layers. The steady state that has now been reached will last for 60 million years, during which the star radiates steadily at the luminosity of 10 000 Suns.

During this short lifetime, the star transforms into helium all the hydrogen whose temperature exceeded 10 million degrees; that is, about all the hydrogen within a central sphere, extending out to one-tenth of the star's radius. In spite of the compression that increases the density of the stellar core, these nuclear reactions reach only a small percentage of the total available hydrogen.

Throughout its life in its 'adult' stage, the star remains practically at the same location on the main sequence, marked 30 $M_\odot$ in Figure 3.1; yet, its radiation rate grows imperceptibly as time goes by. However, after 60 million years, faster changes occur. The core has used up all its fuel and the radiation pressure, which sustains the weight of the outer layers, falls suddenly. The star collapses toward its center, seeking a new steady state. The center reaches 100 million degrees, which brings helium to the ignition temperature for a new nuclear reaction. This time, three helium nuclei fuse into a single carbon (C) nucleus (see Table 3.1). The core continues its contraction until it finds a new steady state near 400 million degrees. At this stage, the temperature has also risen in all the deep layers of the star. In particular, the 10 million degree zone has expanded beyond the helium core. A layer of hydrogen, which until then has not been quite hot enough for ignition, begins to burn into helium.

When the inner layers collapse toward the center, the outer layers undergo an inverse process. Since the production rate of energy has continued to increase, the outer layers inflate because of the higher radiation pressure. The diameter of the star grows by a factor of about 100, surrounding its core by a giant but tenuous atmosphere and this describes the birth of a 'red giant' as well as the creation of a new element in the Universe, carbon. The coalescing of a carbon nucleus with another helium nucleus creates a second new element, oxygen

(O). This synthesis proceeds more slowly and becomes more effective at a higher temperature. Finally, at about 600 million degrees, neon (Ne) and magnesium (Mg) appear.

Compression of the stellar core has another consequence: the electron 'gas', in which the atomic nuclei are immersed, reaches so high a density that the electrons become 'degenerate'. This means that they will not compress any further; the thermostat, which so far had adjusted the temperature and the radiation rate so well, is now thrown out of kilter.

At this time, the carbon nuclear reaction runs out of control, and it triggers a transient jump in the star's luminosity. In step with the nuclear reactions, the stellar core contracts and heats up, so that in turn, carbon, neon and silicon become ignited, producing in particular magnesium and iron (Fe).

Thus, the star has already produced the seven new elements, which, with helium and hydrogen, still constitute 99.9% of the matter in the Universe. The missing 0.1% represents the sum of the other 83 elements of the Periodic Table.

It is important to stress again that neither solids nor liquids have appeared anywhere yet. When I mention iron or magnesium at this stage, I always mean a plasma so hot that there are no neutral atoms but a mixture of atomic nuclei and free electrons. Because the nuclei are 100 000 times smaller than the atoms, they can reach surprisingly high densities without touching. On the other hand, since the electrons are no longer bound to the atoms, it is the compression of these free electrons that sets the density at the star's center. The degeneracy of the electrons is a quantum property that limits the maximum density of the center of the star at 100 tons/cm^3, which is 100 million times that of water.

The series of nuclear fusions that has transformed light atomic nuclei into heavier and heavier ones stops at iron, because iron reaches a maximum nuclear stability. When the nuclear reactions that have led to iron are exhausted, the star tries again to compress, and raises its central temperature; but this time, a new phenomenon occurs. The gamma radiations become so energetic that they can photodissociate an iron nucleus. This nucleus breaks down into 13 helium nuclei, liberates four neutrons and *absorbs heat*; this cools the stellar core and triggers its immediate collapse under the weight of the outer layers. The core collapse sets off a series of spectacular consequences: the star explodes in a fast expanding ball of fire, called a *supernova*, which over several months will shine brighter than a whole galaxy.

How is this possible? Let us first consider the fate of the star's core. It consists of a plasma of iron nuclei and electrons, and reaches 6 solar masses. Its fast collapse makes an '*implosion*' that compresses the electrons so much that they penetrate into the inside of the iron nuclei, and into the new helium nuclei in the process of formation. Each electron combines with a proton to form a neutron (plus a neutrino that escapes from the star), so that all nuclei decay and only a 'purée' of neutrons is left. In a way, one could say that the star's core becomes a single gigantic atomic nucleus of 6 solar masses. Its density is close to 10^{16} g/cm^3 (10 million billion times that of water).

This enormous density curves space–time such that the gravity of 6 solar masses closes space–time upon itself in a black hole of 18 km diameter. The star's core disappears, cut off forever from the rest of the Universe. Only the gravitational potential well survives around the black hole, to remind us that a star is enclosed therein.

However, two fractions escape before the final collapse. They are thrown toward the outer zones of the star by the rebound of the shock wave induced by the resistance to compression of the electron gas. These two escapees are neutrons and neutrinos.

The freed neutrons move with the shock wave through the outer layers, where they collide with other atomic nuclei. With the rapid addition of neutrons, nuclei of carbon, oxygen, silicon, etc. build most of the still missing elements of the Periodic Table. Their fast exothermal nuclear reactions reinforce the shock wave, and throw the outer layers of the star off toward interstellar space at 1000 km/s. This bright, expanding cloud is the supernova proper. It may outshine the brilliance of 100 galaxies for a few days, and remain brighter than a galaxy during the first three or four months of its expansion. Expanding vestiges will continue to be observable for thousands or tens of thousands of years, before they are so scattered as to become undetectable.

The supernova phenomenon is now reasonably well understood. This is illustrated by the correct prediction of all the cosmic abundances of the elements of the Periodic Table (see Figure 4.2). The most convincing predictions relate to elements heavier than iron. Three different processes have been identified. They are rapid neutron capture, slow neutron capture, and (less important) proton capture. *Rapid capture* occurs only at the beginning of the explosion, when the available neutron flux is dense; at that time, a nucleus captures several neutrons consecutively. Intermediate unstable nuclei do not have time to decay before addition of another neutron that transforms them into stable nuclei. In contrast, *slow capture* allows time for the unstable nuclei to decay by beta radioactivity, so that the final stable nucleus is different. Lastly, *proton capture* explains the last gaps not explained by the neutron flux.

Supernova 1987 A, which shone like a beacon for 4 months in the southern sky in the spring of 1987, allowed the first detection of a cosmic neutrino flux not coming from our Sun. The supernova was in the Large Magellanic Cloud, at a distance of some 180 000 light-years. In spite of this huge distance, a flux of 19 neutrinos was detected over 12 seconds. These 12 seconds correspond fairly well to the predicted duration of the gravitational collapse of the star's core. We do not know the date of the collapse, but the neutrino flux travelled almost at the speed of light; after a trip of 180 000 years, the light and the neutrinos arrived almost together, within a few days. This provided valuable information that the rest mass of the neutrino, if it is not nil, must be extremely small; otherwise the neutrino flux would not have been able to travel so fast.

The steady evolution of the less massive stars

Now let us summarize the final fate of two characteristic stars of 10 and 3 solar masses. They evolve much more slowly than a star of 30 solar masses, because their consumption rate of nuclear fuel is much slower. The 10 solar mass star burns up its hydrogen core in 100 million years; the 3 solar mass star does it in 300 million years.

Fate of the 10 solar mass star

After 100 million years on the 'main sequence' described above, the star becomes a red giant that burns its core of helium successively into carbon, oxygen, neon and silicon. Nevertheless, the star is not heavy enough to reach the ignition temperature for silicon into iron. However, its final collapse sets off the same type of shock wave that rebounds into the explosion of a supernova. The major difference lies in the nature of the residual core. It also reaches the density of the atomic nuclei, but its mass is not large enough to curve space–time into a black hole. With only 2 solar masses, a *neutron star* forms and may become observable at the center, when the supernova fades away. The neutron star density stabilizes near 3×10^{15} g/cm^3 (3 billion tons/cm^3!).

The most famous example of such a case is displayed by the Crab Nebula. This nebula is centered exactly at the location where Chinese astronomers observed a supernova in the year AD 1054. It is still expanding rapidly, at the speed of about 1000 km/second. This speed is just what is needed to prove that the Crab Nebula is nothing less than the remnant of a supernova explosion about 950 years ago (Figure 3.2).

The little star at the center of the nebula spins 30 times per second, emitting a luminous flash at each rotation, accompanied by a brief emission of X-rays and radiowaves. This is what is called a *pulsar*. Any pulsar must be a *neutron star* with the density of nuclear matter, for such a spin would shatter a normal star to pieces. Its origin is well understood: the collapse of the central core during the supernova explosion has accelerated its rotation rate. The spinning acceleration results from the conservation of rotational momentum. This is like an ice skater whose arms are initially horizontal and then brought close into the body as she spins.

Fate of the 3 solar mass star

After a much longer stay of 300 million years on the main sequence, this star also becomes a red giant and ends up burning its helium core into carbon. Since its mass is smaller, its central temperature never goes beyond the ignition temperature for carbon. Moreover, its thin and turbulent red giant atmosphere loses mass, since it is progressively lost to a stellar wind.

This wind is much less intense than that caused by a supernova explosion. Nevertheless, it lasts a very long time, and ends up forming a more or less spherical nebula, which envelops the star like a cocoon. The first astronomers to discover such an object called it a 'planetary nebula', not because these nebulas were assumed to give birth to

FIGURE 3.2 The Crab Nebula is the remnant of a supernova explosion observed by the Chinese in AD 1054. It is still expanding at a speed of about 1000 km/second. Located at a distance of some 4000 light-years, its present diameter is about 10 light-years. A faint star at its center is a pulsar which spins 30 times per second; this implies a tiny size and an extraordinary density of 3 billion tons/cm^3, the signature of a hyperdense neutron star. The name 'pulsar' comes from the fact that the star sends out light, radio, and X-ray pulses 30 times per second. The pulsar stems from the central core of the star whose explosion, observed in AD 1054, produced the collapse.

planets, but quite simply because they looked somewhat like the disk of a planet, at least in small telescopes.

When a planetary nebula dissipates and becomes transparent enough, it then reveals the existence of a small hot star at its center. This is a *'white dwarf'*, resulting from the collapse of the star after its helium is completely transformed into carbon. The mass of the white dwarf is never far from 0.7 to 0.8 solar mass. Its final density is set by the maximum possible density of its gas of degenerate electrons, i.e. around 3 or 4 tons/cm^3.

Figure 3.3 shows such a white dwarf at the center of a well-known planetary nebula that has become partially transparent.

The slow evolution of low-mass stars

Let us again consider two typical stars of only 1 solar mass and 0.3 solar mass, respectively. The first remains 10 billion years on the main sequence (this is the case of our Sun, which is still at half-life on the main sequence). Its central temperature rarely goes beyond 12 million degrees, which corresponds to a rate of energy production 10 000 times weaker than that of a star of 30 solar masses. Our Sun is slightly different because it is a second-generation star which already contains carbon, nitrogen, oxygen and metals, but in this discussion, we will ignore these small differences.

After 10 billion years, the core is completely converted into helium and the central temperature rises to 100 million degrees, reaching the ignition temperature when helium becomes carbon. The star is now a modest red giant, 20 times larger and 100 times brighter than our present Sun. This will happen to our Sun in 5 or 6 billion years, and any life on Earth will then be eliminated.

Finally, after a stellar wind dissipates its tenuous atmosphere, the typical 1 solar mass star reveals the dense white dwarf that has progressively formed in its core. This white dwarf has a mass of 0.7 solar mass and a luminosity near 30% of that of our present-day Sun. The degeneracy of its electron gas prevents it from reaching a higher temperature that could ignite new nuclear reactions.

The last case to consider is that of the star of 0.3 solar mass. While on the main sequence, its production rate of energy is 250 times smaller than that of the Sun, which makes it 250 times less luminous and gives it 800 billion years to burn a core three times less massive than that of our Sun. This duration is much longer than the present age of the Universe; hence it will shine quasi-indefinitely into the future, burning just enough hydrogen to hold its center near 10 million degrees.

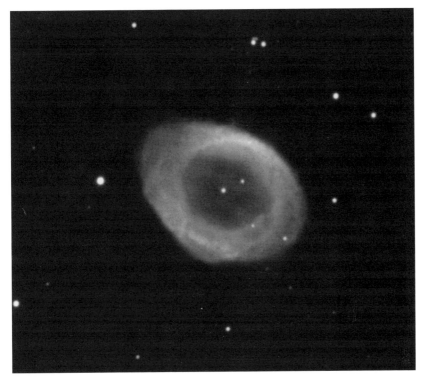

FIGURE 3.3   The planetary nebula in Lyra shows at its center the white
dwarf that remains after the red giant cast off its extended atmosphere
and burnt off all its central helium into carbon. The white dwarf has a
density of 3 tons/cm^3 and was produced by gravitational compression of
the original star's core after the nuclear combustion ended. The other
star off center happens to be on our line of sight, but it is much nearer
and not associated with the planetary nebula.

As for blobs of less than 0.10 solar masses, most have burned only their traces of deuterium. Called brown dwarfs, their glow in the infrared comes mostly from the slow release of their gravitational energy. Their existence is no longer in doubt, because several of them have been observed very recently; they represent an intermediate step between stars and giant planets.

The strange evolution of double stars

The number of double or multiple stars exceeds the number of single stars. Hence the deviations that occur in their common evolution is a feature that affects the final abundances of the chemical elements. These deviations arise from the fact that at particular epochs in their evolution, these stars can exchange matter.

Close binaries are formed by and large by the rupture of an accretion disk that has not kept its axial symmetry. In this case, a density wave in the form of an elongated S emerges in the disk and ends up by condensing into two or more stars.

Let us limit ourselves to the case of two stars that form a close binary pair (Figure 3.4). In close proximity to one another, they revolve fast about their common center of mass. The gravitational potential that separates them can be likened to the saddle shape of a mountain pass that separates two valleys.

Since in general the two stars do not have the same initial mass, the more massive star, which evolves faster, will be the first to reach the red giant stage. Its atmosphere, which then swells quickly, overflows its 'valley' and pours toward the neighboring valley through the 'pass' that separates them. The angular momentum of the gases that overflow causes them to wind in a spiral around the less massive star, where they form a transient accretion disk that ends by increasing the mass of the second star considerably. The first star is likely to end its life soon in a supernova explosion, leaving a hyperdense object such as a neutron star as central residue (it may also disappear in a black hole).

The other star eventually becomes a red giant; it spreads its atmosphere toward a new accretion disk formed this time around the neutron star (or the black hole). Because of the extreme density of this object (whether neutron star or black hole) the potential valley that surrounds it is very deep. The new accretion disk is going to spin at an

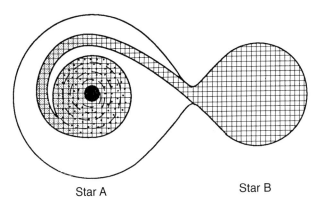

Star A   Star B

FIGURE 3.4   The final phase in the evolution of a close binary. Star A was initially the more massive; hence it evolved faster. It has already exploded in a supernova and left as residue a neutron star, shown in black. Star B has now reached the red giant stage; its atmosphere (shown as cross-hatching) now fills up the whole potential valley and overflows, pouring toward A in a spiral. It forms a transient accretion disk that spins around the neutron star and feeds it. It supplies it with a thin superficial layer of hydrogen, strongly compressed by the extremely large gravity of the neutron star. This compression is large enough for the hydrogen to reach its ignition temperature (10 million degrees) and explode violently, producing the *nova* phenomenon, which can recur every few centuries. The accretion disk is detected by the X-rays it emits. The elongated figure-of-eight curve depicts a contour level that encircles the two potential 'valleys' of the two stars.

extreme speed, and to heat up to very high temperatures, forming an intense source of X-rays around the neutron star or the black hole. The large number of X-ray sources that have been discovered in the Galaxy is explained quite naturally by the large number of close binaries at the end of their evolution.

When the gas from the second accretion disk pours enough hydrogen onto the neutron star, the extreme gravity of that star compresses it so much that its temperature exceeds the hydrogen ignition point. This triggers a runaway reaction, and the superficial layer explodes. This layer is only about one-millionth of a solar mass, but this is enough to light up a 'new' star in the sky. Such a star is called a '*nova*'. It is superficially similar to a supernova, but its brightness is 10 000 or 100 000 times fainter. Although this phenomenon often recurs for the same star after a few hundred years, it seems to play a minimal role in the chemical seeding of interstellar space.

Seeding interstellar space

In the preceding pages, I have described how stars produce the different chemical elements. Of course, molecules cannot form in the stellar core, because it is much too hot. The only chemical elements that played a role in the evolution to greater complexity are those that escaped from the high temperature crucibles where they were formed. Elements escape from stars by two mechanisms: either by an explosive process (novas and supernovas) or by a continuous process (stellar winds).

Small-mass stars have played a negligible role in this process. Their luminosity is too low to induce intense stellar winds, and their lives do not end by any explosive processes. Besides, they live too long, and their evolution has scarcely begun at the present time. In contrast, massive stars play a significant role in seeding the interstellar medium with heavy atoms. First, they have a high luminosity; the pressure generated by their intense radiation develops large stellar winds. Second, they evolve fast and end their lives by explosive phenomena, mostly as supernovas that not only produce the heaviest atoms (beyond iron) but distribute them immediately into the interstellar medium.

The total mass ejected into the interstellar medium is outlined in Figure 3.5 as a function of the stellar masses. It shows the substantial contribution from stars heavier than 10 solar masses. The rate at which each star contributes depends on its lifetime, as summarized in Table 3.2. Thus, only 1 billion years after the onset of thermonuclear reactions in the first stars, all stars of more than 5 solar masses had already scattered into space all the new elements produced by the half billion supernovas occurring in our Galaxy. This seeding of the interstellar medium with all elements heavier than helium was repeated many times in the following billion years, thereby raising the available amount of carbon, nitrogen and oxygen and all heavier elements that astronomers loosely call 'metals' (some are not metals in the sense that chemists define the word).

Table 3.2. *Stellar evolution in brief*

The density of the residual core leads to a paradox, because the black hole density is smaller than that of the neutron star which was not able to make a black hole. Let me emphasize first that these densities are global averages. In particular, the black hole 'density' is obtained by using the Schwarzschild radius for the volume; the mass used is for the residual core that disappeared inside. This is a fictitious density, assuming that Euclidian geometry remains valid. Even so, it is easy to verify that this average 'density' of a black hole *diminishes* in proportion to the inverse square of the mass confined. The density to make a hypothetical black hole with a mass of 1.5 $M_\odot$ would be 8×10^9 tons/cm^3; the neutron star* is not dense enough.

Initial mass of star (in solar masses $M_\odot$)	30	10	3	1	0.3
Luminosity on main sequence (Sun = 1)	10 000	1000	100	1	0.004
Lifetime (main seq.)	0.06	0.10	0.30	10	800
Lifetime (red giant) (in 10^9 year units)	0.01	0.03	0.10	0.30	0.80
Nuclear reactions end with the synthesis of	Iron	Silicon	Oxygen	Carbon	Helium
Final event	Supernova	Supernova	Planetary nebula	Stellar wind	Stellar wind
Mass ejected ($M_\odot$)	24	8.5	2.2	0.3	0.01
Residual core: Nature	Black hole	Neutron star*	White dwarf	White dwarf	White dwarf
Mass ($M_\odot$)	6	1.5	0.8	0.7	0.3
Density (tons/cm^3)	5×10^8	3×10^9	20	10	1

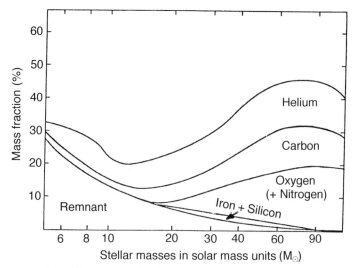

FIGURE 3.5   Mass fraction of the elements synthesized in stellar cores and ejected into interstellar space as a function of the mass of each individual star. Supernovas contribute mostly to the left-hand of the figure, and stellar winds to the right. Toward the right, there is more oxygen than carbon, so that oxides and silicates are formed by stellar winds from supermassive stars. Nitrogen is in parentheses because it is a smaller, late contribution. Toward the left, carbon is dominant for stars of less than 30 solar masses, so that reduced particles of carbon and metallic iron also exist in interstellar space.

The second generation of stars

The slow seeding of the interstellar medium with carbon, nitrogen, oxygen and metals altered the circumstances of stellar formation. With hydrogen and helium only, the first stars ignited hydrogen at 10 million degrees, via the proton–proton reaction.

For stars of the following generations, hydrogen ignition still occurs near 10 million degrees, but combustion goes faster because it is helped near 15 million degrees by the presence of carbon, which acts as a catalyst through the 'carbon cycle'.

In the scheme below, the symbol ν (nu) stands for the neutrino, e^+ for the positive electron, and p for the proton. The neutrino and the positive electron are emitted by the two unstable isotopes ^{13}N and ^{15}O through radioactive transmutation into stable nuclei of carbon and nitrogen, respectively.

$$^{12}C + p \rightarrow {}^{13}N \qquad\qquad \text{then} \qquad {}^{13}N \rightarrow {}^{13}C + \nu + e^+$$

$$^{13}C + p \rightarrow {}^{14}N$$

$$^{14}N + p \rightarrow {}^{15}O \qquad\qquad \text{then} \qquad {}^{15}O \rightarrow {}^{15}N + \nu + e^+$$

$$^{15}N + p \rightarrow {}^{12}C + {}^4He$$

The four reactions on the left are aligned to show their similarity. The energy produced by each reaction has been omitted. The two positive electrons also produce energy, because they are annihilated by colliding with two negative electrons present in the surrounding plasma. This annihilation produces gamma-rays.

The main point of this chain of reactions is that the carbon nucleus, ^{12}C, used at the start is restored at the end, so that what occurs can be summarized by the following generalized reaction:

$$4p \rightarrow {}^4He + 2e^+ + 2\nu$$

Overall, four protons are 'burned' to make one helium nucleus. Traces of carbon are sufficient, because carbon is restored again and again after accelerating the reaction.

I have chosen to examine this process in detail, because it clarifies the way the cosmic relative abundance of nitrogen, as now observed, has been established. The carbon cycle works alternately through nitrogen and oxygen before returning to carbon (this is why it is often called the C-N-O cycle).

However, the first-generation stars do not synthesize nitrogen as easily, because their major reactions are:

$$3({}^4He) \rightarrow {}^{12}C \text{ and } 4({}^4He) \rightarrow {}^{16}O$$

where the synthesis of nitrogen ^{14}N is skipped. Although I have simplified the problem by neglecting minor reactions, it nevertheless remains that the observed abundance of nitrogen would be quite difficult to explain, if the second-generation stars had not solved the problem; the ratio of the reaction rates in the carbon cycle automatically controls the relative abundances of carbon, nitrogen and oxygen.

Other effects of the presence of carbon, nitrogen, and oxygen in a second-generation star are important for astronomers, because they change the star's luminosity somewhat. We will not discuss the matter any further, because it hardly touches on the star's fate or interstellar chemistry. In contrast, the carbon to oxygen ratio in interstellar space led to a fundamental branching of interstellar chemistry. This ratio is altered with the stellar mass. Stars of low mass produce more carbon than oxygen, those of high mass more oxygen than carbon.

During its ejection by a stellar wind or a supernova, the gas soon cools down to a temperature that favors the formation of carbon monoxide (CO), which is very stable and subsists in abundance in interstellar space. After the formation of CO, there remains an excess of either oxygen or carbon in the ejected gas. This means that the expanding bubble is either an oxidizing or a reducing medium. Let us consider the resulting interstellar chemistry.

The chemistry of interstellar clouds

In the Galaxy, the preexisting hydrogen and helium mixed with the millions of gas bubbles coming from supernovas. At the beginning of galactic history, the heaviest stars exploded and therefore ejected more oxygen than carbon (see Table 3.2). After the formation of carbon monoxide, the expanding bubble remained an oxidizer; it now contained in addition to 70% hydrogen and 28% helium, 0.3% carbon monoxide, 0.7% oxygen and 0.1% nitrogen, 0.1% iron, 0.05% magnesium and 0.05% silicon.

But an amount as small as 0.2% oxygen is enough to oxidize all metals and the residual 0.5% oxygen formed water vapor, which became and remained the most abundant triatomic molecule in the Universe (the most abundant diatomic molecule is carbon monoxide). As the temperature dropped, metallic oxides made silicates, which condensed into solid ultramicroscopic grains that water vapor soon covered with frost (liquid water cannot exist in the vacuum of space).

What has occurred merits some attention. For the first time in the history of the Universe, solid matter had appeared. Earlier, when the plasma of the Big Bang cooled down, it formed gaseous hydrogen and helium only, and the mixture became clusters of galaxies, then galaxies. When stars formed within the galaxies, neutral atoms of hydrogen and helium were first ionized before nuclear reactions took place in the crucibles of the stellar cores. When expelled into space, the new elements captured electrons and became neutral again, before forming the first molecules in interstellar space. At last, solid matter could form what we now call interstellar dust. These dust grains were, however, much smaller than what we call dust on the Earth; they look more like wispy smoke.

Massive stars made oxidizing expanding bubbles, in which fine silicate grains were covered by a layer of frost. In contrast, less massive stars made reducing expanding bubbles, where there were no silicate grains and no water. The dust grains are a mixture of reduced carbon grains, in the form of microscopic particles of graphite, diamond and metallic iron. Later, in the second-generation stars in which nitrogen was already abundant, the second- and third-level diatomic molecules would probably be CN and CH. In these reducing bubbles, the first triatomic molecule would not be water, but hydrocyanic acid (HCN). Abundant carbon and superabundant hydrogen formed grains of metallic carbides and polyaromatic hydrocarbons (PAH). The PAH are the stablest of hydrocarbons that can occur in those conditions.

Throughout billions of years, several billions of these huge bubbles formed, in particular along the spiral arms of the Galaxy, where massive stars exploded so fast that they were still close to their birthplace. Since the available mass was distributed equally, the smaller stars were in proportion more numerous. They contributed significantly to the reducing bubbles, in particular by their production of innumerable 'planetary nebulas'.

So, countless bubbles were scattered equally everywhere, from extreme reducing to extreme oxidizing conditions. Because of turbulent motions in the gas, combined with the differential rotation of the galactic arms, these bubbles mixed and interpenetrated, while remaining very far from thermochemical equilibrium; the usual chemical reactions were quenched by the severe interstellar cold.

Unusual chemical reactions were, however, going to take place. Near absolute zero, they depended on rapid chemical kinetics that involve charge exchange between ions and preexisting molecules. The prevailing ions were those of hydrogen and helium, because of the abundance of their neutral atoms. Hot stars radiate intensely in the ultraviolet, thus ionizing hydrogen and helium whose ions provide the energy source for the fast charge-exchange reactions.

The interstellar clouds that built up along the spiral arms of the Galaxy hold at the present time only 10% of its mass. They have sizes of 10 to 100 light-years, and a large range of densities, from a few particles per cm^3 to some millions of particles per cm^3. Regions around hot stars are not favorable to the building of large molecules, because there are too many H and He ions. The densest clouds, being opaque, cool off by radiating in the infrared, and the opacity protects new molecules. The temperature varies by location; far from ionization sources and hot stars, it goes from 80 K down to 10 K.

The density of the clouds varies a great deal, and astronomers try to classify them as 'diffuse clouds' or 'dense clouds'. The opacity of the densest clouds protects a complex interstellar chemistry; in these clouds, the most abundant molecules are molecular hydrogen (H_2) and carbon monoxide (CO). At a much lower level of abundance comes a long list of more than 100 different molecules that have been identified with astonishing accuracy by the wavelengths of their rotational lines as detected by radiotelescopes. However, the abundance ratios vary from one cloud to another, and remain somewhat uncertain. A list of these abundances, by orders of magnitude, in parts per million (ppm) of the abundance of *atomic* hydrogen (molecular hydrogen is less abundant) is given in Table 3.3.

With H_2 (whose abundance varies), I have named the 14 most abundant molecules in the dense clouds of our Galaxy, and presumably in all galaxies. In particular, we can recognize in the first two rows CO (carbon monoxide), N_2 (nitrogen), H_2O (water vapor), H_2CO (formaldehyde), HCN (hydrocyanic acid), as well as familiar molecules like NH_3 (ammonia) and CO_2 (carbon dioxide) and unfamiliar ones like HNC. Let us stop here because, further down in the degree of abundance there is a list of more than 80 molecules, of which more than 60 are organic molecules, all precisely identified by their microwave emission spectra recorded by radiotelescopes. Not a year goes by without the addition of one or two more identifications, mostly of organic molecules.

A quantitative theory of the formation of interstellar molecules has emerged in the last decade. It allows some predictions that are verifiable. However, it has come up against two obstacles: the lack of kinetic constants for ion–molecular reaction rates, and the lack of thermodynamic data for polyatomic ions. Otherwise the fundamental mechanisms for the formation of large molecules seem well understood. Interstellar chemistry depends only on chemical kinetics, because the results remain always far from thermochemical equilibrium. On the other hand, all reactions are necessarily exothermic, because entropy (see Glossary) vanishes in the vicinity of absolute zero.

Table 3.3. *Molecular abundances in interstellar space*

About 10 ppm:	CO	N_2		H_2O	H_2CO	
About 1 ppm:	HCN	HNC		NH_3	CO_2	CH_3OH
About 0.1 ppm:	CH_4	CH_3-O-CH_3				
About 0.01 ppm:	SO_2	SO				

Since the density of what astronomers call a 'dense' cloud corresponds to a high vacuum in our terrestrial laboratories, there are never triple collisions between atoms and molecules, except on the surface of solid grains, where molecules can stick and stay longer. This is the reason why even very small solid grains play an important role as catalysts for certain reactions.

Finally there is no doubt that the major source of energy is the ionization of H and He by ultraviolet light from stars. In particular, He^+ is very effective in producing cascades of ionizations by collisions with molecules, because its ionization potential (24.6 eV) is larger than that of hydrogen (13.6 eV).

Before leaving interstellar chemistry, we must stress that the 'prebiotic' molecules, i.e. molecules required to start life on Earth, exist in abundance everywhere in space. The three most important prebiotic molecules are probably water (H_2O), formaldehyde (H_2CO) and hydrocyanic acid (HCN). These crucial prebiotic molecules reach high abundances in our Galaxy and presumably everywhere else in other galaxies. The evolutionary route that led to life seems to have taken the way with the fewest obstacles and chosen the most abundant construction materials available.

THE FORMATION OF THE PLANETS

Hence there are innumerable suns and an infinite number of earths turning around these suns in the same way as the seven earths that we see turn around the sun which is near us.

Giordano Bruno, *De l'infinito universo e mundi*, 1584

The Universe grows old

Enormous stretches of time have elapsed since the Big Bang. The twilight of the first million years has been transformed into near complete darkness. The fossil radiation, a relic of the Big Bang, that was still dimly lighting the large cooling masses of gas, has diluted and shifted to the infrared, because the expansion of space goes on. This radiation soon becomes completely invisible. In the opaque night of the first billion years, the masses of gas become more and more patchy. This is because density fluctuations increased, and the aggregates of gas were more and more separated, first into superclusters, then clusters of galaxies, and finally into galaxies.

The 'timeless night' probably ends during the second billion years, because the quasars light up the central clusters of many galaxies. Their dazzling light hides the simultaneous appearance of many small bright dots that studded the galactic halos. The stars have just lit up, in the globular clusters and in the large central cluster of many galaxies.

The galaxies each evolve somewhat differently. They display a large diversity in sizes, and in angular momenta, which comes from the turbulence in the gas masses. Most small galaxies form elliptic systems that seem to have undergone an accelerated evolution; all their stars are already very old, so that red dominates their color. On the other hand, they seem not to have produced many second-generation stars, because the spectra of extant stars show a notable lack of metals. Finally, these small galaxies do not seem to have large dust clouds. Other galaxies may look very different, as for instance those with an irregular form that reveals an absence of rotation.

Space was much smaller in the early Universe, hence it is not surprising that there were more frequent galactic collisions. Many of the present-day large galaxies, including those with a spiral structure, may have emerged from these early collisions. In most cases, these large galaxies separated their masses of gas into two components: a dense central lenticular cluster that looks a little like a small elliptical galaxy, and a huge disk of gas that has become very flattened through the billions of years. The disk rotates about the central cluster with a speed that increases toward the center, winding out spiral waves of density that are amplified by the action of gravitation.

These large galaxies are most often surrounded by a halo that is more or less spherical, where hundreds of globular clusters are scattered. The globular clusters, although much smaller than the elliptical galaxies, must have evolved in the same way, because the metal content of their stars is on average 100 times less than that of later stars, and in the clusters no trace of dust clouds is to be seen. Just like the small elliptical galaxies, the globular clusters must have exhausted the available gas so fast that they were unable to produce second-generation stars.

In contrast, the galactic spiral arms live an active life, because density waves stimulate the constant formation of new stars. As a matter of fact, spiral arms remain studded with bright blue specks, typical of young stars. Hence it is in the spiral arms that new generations of stars follow one another and intermix. Since stars of lesser mass have long lives, they have ample time to leave the spiral arm and survive elsewhere in the disk. In contrast, hundreds of generations of very massive stars follow one another and explode into supernovas without leaving the spiral arms.

Hence, in the spiral arms the supernovas also enrich countless galactic clouds with microscopic dust grains and organic molecules. In the dense clouds, the microscopic grains of silicates of iron and magnesium are cold; they are covered with a frost made up not only of water snow, but also of snows of carbon dioxide (CO_2), formaldehyde (H_2CO) and other organic compounds containing nitrogen and oxygen. They are intermingled with particles from elsewhere, mostly graphite, PAH and iron, but also heterocyclic systems whose molecular rings contain carbon and nitrogen, for example purines and pyrimidines whose role in the emergence of life will soon become apparent.

From these billions of evolving galaxies, we now turn our interest to a particular one – ours. It is very large, but this feature does not make it conspicuous among millions of other giant galaxies. It probably contains at least 500 billion times the mass of our Sun. A large part of its mass is far from the center and invisible, apparently formed of dark matter (extinct stars, non-luminous gas, brown dwarfs, etc.), matter that is detected through its influence on the orbits of visible stars near its periphery.

After evolving over some 10 billion years, the spiral arms in our Galaxy still held some 10% or 15% of gaseous reserves available to make stars. It was in one of these spiral arms of the Galaxy, at about 30 000 light-years from its central cluster, that our Solar System was born fewer than 5 billion years ago, and about 8 billion years after the Big Bang. Since we wish to understand the chemical composition of the planets that were going to appear, we must first examine the relative abundances of the chemical elements that were available in the giant molecular cloud which would eventually form the Solar System.

Cosmic abundance of the elements

All the chemical elements exist in nearly the same proportions in the Sun and in all its neighboring stars. These proportions do not seem to be essentially different from those of the nearby interstellar medium, at least when the dust is also taken into account.

Since the whole of the Solar System had to begin from a 'primitive mixture' identical to that which formed the Sun, it has been encouraging to discover that there are still objects in the Solar System that carry the signature of the abundances of the elements, in particular the primitive meteorites called *chondrites*. We will return to chondrites later on, but it is worth noting that they have become the source of the best determinations of cosmic abundances for some 60 or so elements that are not very volatile (called 'metals' by astronomers). These results have confirmed, with more accuracy, all the metallic abundances found in the spectrum of the Sun. However, the Sun is needed to fix the abundances of gases such as hydrogen, helium, nitrogen and oxygen. It is clear that the results apply only to 'young' stars such as the Sun and stars in its vicinity that are all second-generation stars. 'Second generation' here is merely a figure of speech, since there have already been many generations of massive stars that have seeded the interstellar medium in our galactic arm over some 12 billion years.

On the other hand, very old stars can still be observed in globular clusters (see Appendix D). There are also similar stars in the galactic halo as well as in the central cluster; their spectra show that their metal content is 100 to 1000 times less and therefore they are very old. However, the traces of 'metals' they contain reveal that they have lost the purity of the primitive stars.

We recall that each of the elements exists as several isotopes, whose chemical properties are identical, but whose atomic masses differ. The isotopes come from the varying number of neutrons present in the atomic nuclei. It is the number of protons that defines the chemical properties of an element, so that the isotopes are not distinct chemically. Figure 4.1 shows the number of protons and neutrons in all the known stable atomic nuclei, as well as in radioactive nuclei with long lifetimes (e.g. uranium). In the lightest atoms, there is about the same number of protons and neutrons, while in the heavier atoms there are always more neutrons than protons.

Although all isotopes are important, to simplify the discussion we will consider only the most abundant isotope for each species of element existing in nature. In Figure 4.2, the most important isotopes have been classified by their increasing number of protons. Their relative cosmic abundances are given as the number of atoms per million atoms of silicon, and are plotted logarithmically. The abundances fall very steeply towards the bottom of the diagram; in the Solar System, all elements beyond zinc (symbol Zn, atomic number 40) are extraordinarily scarce, since they are more than 10 000 times less abundant than silicon.

We cannot, however, directly compare the cosmic abundances with those of the elements present on the Earth. The reason is that the Earth has undergone thermal differentiation: its crust is separated from a hot magma, so the proportions of the elements are very remote

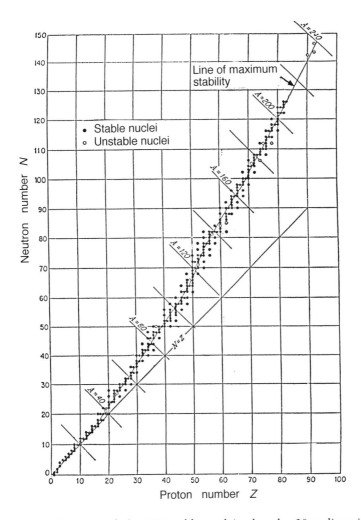

FIGURE 4.1   Diagram of the 270 stable nuclei, plus the 20 radioactive
nuclei with a half-life greater than 100 million years. The atomic
number Z is the number of protons in each nucleus. The atomic mass A
is the total number of protons and neutrons, so that the total number of
neutrons N, plotted as ordinates, is given by $A-Z$. Note that, up to
atomic mass A = 40 of neon, the nuclei contain about the same number
of protons and neutrons. For heavier nuclei, the line of maximum
stability deviates progressively toward a larger number of neutrons than
protons. The flux of neutrons released in supernova explosions allows
the synthesis of those heavy elements that are stable enough to survive
for a long time.

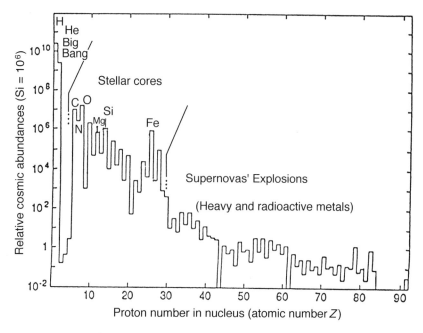

FIGURE 4.2   Relative cosmic abundances of the chemical elements
classified as a function of their atomic number Z (this number is also the
classification number in the Periodic Table). The abundances are given
as the number of atoms relative to 10^6 atoms of silicon. These are the
abundances found in the Solar System and in nearby stars (as young as
the Sun). The abundances of hydrogen and helium date from the Big
Bang; the elements from carbon (Z = 6) to beyond the iron peak (Z =
26) come from the nuclear synthesis in stellar cores, while the elements
beyond the iron peak arise from supernova explosions that seeded the
molecular cloud that eventually formed the Solar System.

from the cosmic abundances. Tectonic processes have concentrated
some metals at certain places; in turn, some gases considered 'rare' on
Earth are very abundant in the cosmos. The best example is probably
helium, which is second, after hydrogen, in the order of cosmic abundances.

What is remarkable in the observed cosmic abundances is the fact
that we understand the origin of each of the peaks in Figure 4.2. They
separate naturally into three groups. First, there is the major peak of H
and He, the first two elements, separated by a deep trough from the

sequence that runs through atomic numbers 5 to 30, followed by an unmistakable fall off for the peaks beyond 30.

The first group dates from the Big Bang; its origin was explained in Figure 2.1. The second group comes from the successive nuclear syntheses in the cores of stars. The third group was synthesized during supernova explosions; in particular, neutrons freed by the exploding stars provided the necessary excess to build up the heaviest nuclei. The obligatory neutron excess is shown in Figure 4.1: the line of maximum stability goes much higher than the 45° diagonal. The source of neutrons in the exploding star is the photodissociation of iron (see Table 3.1).

Instead of numbers of atoms, if we now consider their *mass ratios*, we see that 99.97% of the existing matter in the Solar System is supplied by only 10 of the 92 elements shown in Figure 4.2. The sum of the other 82 elements represents only 0.03%. Let us talk in more detail about those 10 major elements.

Hydrogen and helium together constitute 98.1% of the total mass of matter. Carbon, nitrogen and oxygen are the products of the very first syntheses of new elements in the cores of stars; they make up a total of almost 1.4% of matter. Then the three most abundant metals (iron, magnesium and silicon), with sulfur and neon, complete the list of the ten most abundant elements, and together add only almost 0.5% of the total mass. Let me emphasize that iron is the most abundant metal because it forms the most stable final stage of the thermonuclear syntheses in the cores of stars.

Finally, the numerous 'rare earths' of the chemists – the heavy metals (e.g. gold, platinum, lead, silver), and *all* natural radioactive metals (e.g. radium and uranium) – were synthesized thanks to an excess of available neutrons from supernova explosions. The steady decay of all radioactive elements allows an assessment of the average age of all supernova explosions that have contributed to the elemental abundances in the Solar System and, by extrapolation, of the age of our Galaxy (see Appendix D).

To what degree are the elemental abundances determined for the Solar System also representative of those in the giant galactic cloud from which the Solar System and nearby stars were born? Was there any mixing, and at what level do inhomogeneities continue to exist? For instance, some nearby stars show somewhat less carbon content than the Sun. However, the best indications come from two sources: one of them is the abundant information that chemists have amassed on chondrites; the other comes from the analysis of ultramicroscopic dust grains that were freed by Halley's comet and intercepted in 1986 by the Giotto Space Probe launched by the European Space Agency.

The chondrites

Chondrites are primitive stony meteorites that come from the asteroid belt. They are the result of collisions between small asteroids. Some fragments are ejected into space, and their trajectories end up in the Earth's atmosphere. Those retrieved on the ground are usually now in museums. Chondrites can be distinguished from other meteorites, because they constitute *undifferentiated* samples of the primitive material that formed the Solar System. In larger asteroids as well as in the planets, primitive matter has often been differentiated and transformed by processes involving gravity and heat.

Analysis of chondrites reveals that they have never been heated, even to moderately high temperatures, and that they result from *dust*, compressed slightly by the extraordinarily weak gravity (typically, less than 1% of the Earth's gravity) of the very small asteroids from which they came. This fact sets the size of these asteroids at less than 100 km diameter.

Quantitative chemical analysis of the elements present in chondrites shows that their relative abundances are exactly the same as all those known for the Sun. They yield all the abundances given in Figure 4.2, except for some of the more volatile elements, which were evidently not retained by the very weak gravity of their original asteroid. Chondrites have been divided into four classes according to their oxidation pattern: completely oxidized carbonaceous chondrites, incompletely oxidized carbonaceous chondrites, 'ordinary' chondrites (still somewhat oxidized), and totally reduced *enstatite* chondrites. It is obvious that not all chondrites are equally primitive. Some have been influenced by minor chemical modifications due to their thermal history in the asteroid belt. The carbonaceous chondrites are indisputably the most primitive, in particular those which are totally oxidized, because they also contain volatile elements in the highest proportions.

Those carbonaceous chondrites consist of 85% to 90% hydrated silicates of iron and magnesium, plus 3% to 4% carbon, and a small percentage of sulfur and of uncombined water (in addition to the hydration water of the silicates). They have the appearance of quite light pebbles, with a density in the range 2.2 to 2.7 g/cm^3 (which compares with 3.3 to 3.5 for ordinary terrestrial pebbles). Microscopically, they show well-defined mineral grains, for example particles of serpentine $Mg_3Si_2O_5(OH)_4$. Serpentine is a hydrated silicate of magnesium well known in geology. There are also grains of other silicates of iron and magnesium in different proportions. However, if one considers not just a particular grain but a large number of them, they always contain an average of 95 to 100 parts of magnesium to 100 parts of iron, and this is the proportion found in the Sun (in Figure 4.2, the Fe peak is slightly higher than that of Mg).

Such a comparison has been done for more than 30 metals, and led to the belief that the Sun and the chondrites came from a common source of matter. In addition, the history of chondrites and in particular of carbonaceous chondrites must have been very uneventful, otherwise the proportions of all the elements would not have been so well preserved.

Their submicroscopic examination has led to more surprising revelations. The silicate matrix is made up of grains about 0.1 μm in diameter, compressed just enough to be fused together. This is surprising because the grains are not in chemical equilibrium with one another. Grains that are totally oxidized touch grains of metallic iron (i.e. totally reduced); refractory grains are in contact with volatile grains; some grains have been modified by liquid water, and they touch totally anhydrous grains.

Occasionally, the matrix imprisons larger objects, of about 1 mm or more. Called 'chondrules', these little balls of silicate show signs of temporary fusion: they are droplets that have cooled down. Sometimes, there are also large refractory grains containing minerals rich in calcium and aluminum, which form only at high temperature, typically more than 1600 K.

This heterogeneous composition seems to imply that the grains have different origins and came from different places, eventually being assembled by a cold process suggestive of sedimentation. We easily understand the origin of terrestrial sedimentary rocks. In a river, when the turbulence of water subsides, the sand settles to the bottom of the channel. The process collects grains of various origins – quartz, feldspar, mica, calcite and silicates. These grains have been brought by streams flowing from nearby hills or distant mountains. After long geological periods, these grains can aggregate into a single rock, but we can still recognize their origins easily. On an asteroid, however, there is no atmosphere, hence no liquid water; there is not enough gravity either to make sediments.

There was a time, before the planets existed, when a process comparable to sedimentation separated gas and dust. This happened after the formation of the accretion disk that surrounded the early Sun. First, the collapse of the interstellar cloud which fed the disk sustained turbulence in the gas, which kept the dust in suspension. When the collapse died away, the gas turbulence in the disk subsided and the interstellar dust could settle down, falling under the influence of gravity into the equatorial plane of the disk. In this plane, the dust formed huge thin rings, comparable to the rings of Saturn, although much larger.

What we see in the matrix of carbonaceous chondrites, therefore, are still recognizable interstellar grains, and what is in them is of great interest. In most of the silicate grains, there are 1% to 5% carbon compounds. Most of it is in the form of an insoluble polymer of high molecular weight, difficult to analyze because of its insolubility, but which contains hydrogen, oxygen, nitrogen and sulfur. The rest of the organic matter present has been studied in much detail, in particular in the carbonaceous chondrite that fell in 1969 at Murchison in Australia.

A total of 411 different organic compounds were identified in the soluble fraction. There were in particular (in parts per million (ppm)) 44 carboxylic acids (total 350 ppm); 227 aliphatic acids and aromatic carbohydrates (total 60 ppm); 74 different amino acids (total 60 ppm); 12 amines and amides (total 60 ppm); 17 alcohols, aldehydes and ketones (total 40 ppm); 32 heterocyclics containing nitrogen (total 7 ppm) including 4 purines (adenine, guanine, xanthine and hypoxanthine) and 1 pyrimidine (uracil).

It is not necessary to go into further detail. Let me just draw attention to the organic compounds that could have played a role in the emergence of life on the Earth. Among the 74 amino acids found, 8 are used in the proteins of living organisms; these are glycine, alanine, valine, leucine, isoleucine, proline, aspartic acid and glutamic acid.

Table 4.1. *Analysis of cometary dust*

Organic fraction	(%)	Inorganic fraction	(%)
Hydrocarbons	16.0	Silicates	51.5
H, C + O	5.2	Troilite (FeS)	6.0
H, C + N	4.5	Graphite (C)	3.0
H, C + S	1.8	Sulfur (S)	1.0
Water	5.5	Water	5.5
Total	33.0	Total	67.0

Note:
In the organic fraction, the notation H, C + O indicates the fraction of
organic compounds that contain oxygen (and similarly for nitrogen (N) and
for sulfur (S)). It was not possible to analyze these fractions in more detail.

Eleven others are used in less common biological systems. As for the
remaining amino acids, they have no known connection with biology.
However, all purines found, as well as pyrimidine, are used in biological systems, in particular in ribonucleic acid (RNA) and in deoxyribonucleic acid (DNA), where they play an essential role as the
'letters' of the genetic code.

Cometary dust

The only analyses of cometary dust have come from the mass spectrometer carried on board the Giotto space probe; they cover the crucial
minutes when this probe crossed the dust tail of Halley's comet, in the
immediate vicinity of its nucleus. However remarkable, these results
cannot be compared to the richness of analyses carried out on numerous chondrites during past decades of patient work by numerous chemists. Let us first give the average composition of this cometary dust.

In the inorganic fraction given in Table 4.1, it may be worth men-
tioning that water is assumed to be the hydration water of the silicates.

Each dust grain was analyzed separately by mass spectrometry. This
method gave an indication of surprisingly large differences among
individual grains. By and large, the grains could be classified into four
different families, as follows:

(A) 37 carbon-rich grains (average 54% C, 12% O, 10% metals)
(B) 18 oxygen-rich grains (average 54% O, 10% C, 15% metals)

(C) 10 magnesium- and silicon-rich grains (average 81% metals, but only 5% Fe, 2% C, 10% O)

(D) 11 iron-rich grains (average 33% Fe, 9% Mg, 5% Si, 8% C, 4% O)

Since the grains show such large differences, it is quite surprising that they fall into families that are so distinct. We have to believe that we are faced with individual stellar grains that were born in at least four different stars:

(A) A star rich in carbon (hence reducing)
(B) A star rich in oxygen (hence oxidizing)
(C) A star rich in magnesium and silicon, somewhat oxidizing ($C/O = 0.2$)
(D) A star rich in iron and somewhat reducing ($C/O = 2$)

This interpretation is confirmed by important anomalies in the isotopic ratios of several major elements that are not within the scope of this book.

Organic molecules that include purines and pyrimidines have also been identified in the grains of Halley's comet. We have already emphasized the importance of these prebiotic molecules also found in chondrites. There is not much doubt that many prebiotic molecules have appeared in interstellar space, have accreted in the dust present in the accretion disk surrounding the early Sun, and have accumulated and been kept in pristine condition in comets.

Variety of interstellar chemistry

The microscopic evidence found in the most primitive bodies extant in the Solar System, namely the carbonaceous chondrites and the comets, demonstrates that all bodies of the Solar System were originally formed by a disparate assemblage of oxidizing and reducing interstellar grains.

We can assume that they came from different stellar sources and were initially stored for a long time in the deep cold of an interstellar cloud. This cloud collapsed into what was going to become a large accretion disk surrounding the early Sun, and the grains sedimented into the mid-equatorial plane of this disk, where their accretion led to the planets.

The heat coming from the accretion process brought the grains a little closer to thermochemical equilibrium. The oxidizing or reducing conditions are determined somewhat by temperature, but mostly by the overall ratio of oxygen to available carbon.

The reason derives from the stability of carbon monoxide (CO) at relatively high temperatures; at lower temperatures, the CO molecule remains metastable because its conversion to methane (CH_4) is inhibited by the slow rates of the possible reactions, despite the large overabundance of hydrogen (H_2). Since the formation of CO uses as much oxygen as it does carbon, there will remain an excess of either carbon or oxygen, depending on whether the overall C/O ratio is larger or smaller than 1.

But we know that there is three times more oxygen than carbon in the Sun. Since the whole planetary system has the same origin, its primordial overall conditions were oxidizing. This is indeed what is observed in the carbonaceous chondrites and in Halley's comet, in spite of the preservation of large disparities.

We can conclude that at the time of the formation of the Solar System, many stars of the first generations had already mixed numerous organic compounds (containing C, N, O and S) with primitive hydrogen and helium. The dust showed a predominance of silicates, and a lesser amount of reducing grains such as metallic iron or graphite and polymers of aromatic hydrocarbons. A layer of frost coated the grains, holding not only water snow in abundance, but also snows of volatile organic compounds, including in particular all the molecules propitious to the emergence of life.

A few billion years later, we now observe another giant molecular cloud undergoing the same events, namely the molecular cloud that covers a good part of the constellation of Orion. This giant cloud is located in the spiral arm of the galaxy nearest ours (see Figure 4.3). It includes thousands of unusually young stars. Several hundreds of them were discovered, in 1993, to display circumstellar dust rings, obvious remnants of their accretion disk. It is not unreasonable to think that the Orion nebula contains hundreds of planetary systems in the process of being formed.

FIGURE 4.3   A fragment of the giant molecular cloud in the constellation of Orion, which is in the process of condensing into millions of new stars. The four central stars are very massive stars that formed first and burn quickly. Only that part of the giant molecular cloud which is in their vicinity is lit by their ultraviolet light. The biggest part of the cloud is opaque and dark; it covers a large part of the constellation of Orion. The hydrogen molecule is not detectable by radiotelescopes, so CO is used to set the limits of the cloud. It must contain a mass of gas of about 1 million solar masses. Several hundred stars in the process of formation were discovered in the cloud in 1993. Most are surrounded by rings of dust, apparently the remnants of their accretion disk that might soon form planets. These observations were the first to suggest that planetary systems like ours must be abundant in the Universe.

Gravitational collapse

Let us go back to about 5 billion years ago, in the spiral arm of the Galaxy where the Sun is going to form. We see a giant molecular cloud quite similar to the one that appeared 5 billion years later in the constellation Orion. It contains a mass of gas of about 1 million solar masses, and is opaque enough to have cooled while radiating in the infrared, until its center went down to a very low temperature, perhaps 10 K. It consists of 70% hydrogen, 27% helium, 1% oxygen, 0.3% carbon and 0.1% nitrogen. By charge exchange with ionized atoms, these elements have already formed a large number of molecules that are mostly organic. Thanks to the cloud's opacity, they are protected from the ultraviolet of distant stars. The cloud also holds 1.5% dust by mass: three-quarters of this are grains of silicate of iron and magnesium, covered with water frost and snows of volatile organic matter; one-quarter of the dust is made of grains of carbon, polyaromatic hydrocarbons and metallic iron. A part of the available carbon and oxygen has remained in the gas, in the form of 0.1% carbon monoxide. Within the opaque core of the cloud, there are no longer many ions – perhaps one ion per ten million molecules – but, together with the free electrons, they constitute a plasma that reacts to magnetic forces.

This molecular cloud has an irregular structure that shows some turbulence. The extant blobs have many sizes that show the scale invariance of a fractal. Each blob has a slightly denser center toward which it tends to contract by virtue of its own gravity. However, its weight is supported not only by the turbulent motions of the gas, but also by the magnetic field bound to the plasma. But the magnetic field is not strong enough to withstand gravity indefinitely. It slowly diffuses during the collapse of the blobs toward their center, slowing them down over a few million years. Sooner or later the blobs collapse into much smaller objects, the more massive first, forming massive stars of 10, 30 or even 60 solar masses.

These stars form over some hundreds of thousands of years through the mechanism of the accretion disk. Because they are very massive, their nuclear fuel will be used up fast after a few million years, and they will explode into scores of supernovas. They create hot, oxidizing bubbles whose expansion propagates very rapidly, as a shock wave to the very midst of the molecular cloud. The shock wave compresses less dense blobs nearby and helps to trigger their gravitational collapse; hence some hundreds of less massive stars are going to be created almost simultaneously.

Among these, there is a nodule scarcely larger than 1.5 or 2 solar masses. It is the one we will follow in detail through all its evolution, because it is going to form our Solar System.

The accretion disk of the Sun

The nebular nodule that is going to form the Sun and the planets collapses toward its center of mass (Figure 4.4). When this occurs, the distances between all parts of the nodule decrease by a huge factor, of the order of 100 000. The central part of the nodule collapses first; it falls in faster and faster. But the turbulent motions of the gas have also an imperceptible initial rotational velocity. It has what is called 'angular momentum', defined by the product $V \times R$, V being the component of velocity *perpendicular* to the direction of the center, and R the distance to the center of mass. Mechanics tells us that, without intervention from outside, angular momentum remains constant. This implies that, as the distance R becomes smaller, the velocity V increases in proportion. While the blob contracts by a factor of 10 000, the rotational velocity around the center increases by the same factor.

This whirling prevents the nodule from collapsing completely toward the center. Only the central fraction, perhaps 5% to 10% of 1 solar mass, takes a globular shape that compresses and contracts. This globule heats up and stabilizes while radiating as a big red star. The rest of the nodule flattens out into a large disk that whirls around the proto-Sun. The slow collapse of the periphery of the nebular nodule is going to last at least 100 000 years, and its fall will be stopped by a shock wave developing at the surface of the disk.

The accretion disk consists of 98.5% gas and 1.5% dust (by mass). The dust is kept in suspension by the gas turbulence. This turbulence is maintained, throughout the collapse, by the transformation of half of the kinetic energy from the infall into turbulent energy, which will end up as heat, raising the gas temperature. The other half of the kinetic energy from the infall is conserved as the kinetic energy of rotation of the whirling disk.

The turbulent viscosity of the gas causes friction between adjacent

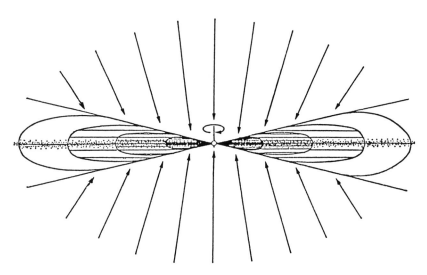

FIGURE 4.4 Accretion disk of the Solar System, in cross-section perpendicular to the equatorial plane. The future Sun is at the center. The arrows represent nebular matter falling in from afar; it is stopped in its fall by a shock wave in the shape of a very open cone. This falling matter keeps turbulence going and heats the disk. Any collapse of this kind transforms the potential energy of gravitation, half into kinetic energy of rotation and half into turbulent motion that eventually turns into heat. The closer matter goes to the center, the more it heats up and the more it whirls. Because of the turbulent viscosity of the disk, the inner zones spiral in toward the center and add to the Sun's mass, while the outer zones enlarge and swerve out from the center. When the fall of nebular matter ends, the Sun reaches its final mass, turbulence subsides and the dust settles out toward the equatorial plane of the disk. This type of sedimentation process makes thin dust rings comparable to Saturn's rings, although much wider.

zones, because the zones turn at different angular speeds, in accordance with Kepler's laws. Just as for accretion disks, matter in the inner zones is slowed down and drawn in a spiral nearer to the central mass to which it finally accretes. Matter in the outside zones is accelerated and broadens the rings, carrying away the excess angular momentum outward. Thus, within 100 000 years, the Sun has slowly reached its final mass.

At that time, the nebular nodule has exhausted its material and has finished collapsing; hence the turbulence subsides in the disk. There comes a time when the gas turbulence cannot support the weight of the dust any more, and the dust settles out as it falls toward the central plane of the disk. We recognize here the dust sedimentation from gas which left fossil traces in the primitive bodies such as carbonaceous chondrites and comets. It is remarkable that this separation of the solid phase from the gaseous phase happens very quickly, probably within a few thousand years.

This is therefore the sedimentation epoch which best defines the 'age zero' of the Solar System, and which different determinations of the radioactive age of chondrites measure with precision: it occurred 4.56 billion years ago.

Temperature distribution in the disk

As I have mentioned above, when matter falls toward a center of attraction, the gravitational energy is transformed half into kinetic energy and half into heat: astronomers refer to this as the 'virial' theorem. The deeper it goes into the gravitational 'well' of the central mass, the more the accretion disk speeds up and the more it is heated. Notably, it is because the gravitational well is abnormally deep (i.e. the gravitational potential is extremely intense) in the vicinity of hyperdense objects that their accretion disks reach outstanding temperatures. This explains the X-rays emitted around neutron stars as well as the astonishing luminosity of the quasars.

The situation was not fundamentally different in the accretion disk that formed the Solar System, except that the disk never reached extreme temperatures because the Sun is a star of small mass and moderate density. The most characteristic property of an accretion disk is its radial distribution of temperature. The gravitational potential energy is a negative energy that varies as the inverse of the distance to the central mass. Since half of the kinetic energy of that matter which falls into the potential 'well' is transformed eventually into heat, the temperature of the disk will also vary as the inverse of the distance to the center.

The disk may be opaque, since it can be thick and contains dust. If so, its surface behaves like the photosphere of the Sun: the hotter it is, the more it radiates and cools, which alters the temperature distribution on the outside surface of the disk. On the other hand, in the median plane of the disk, the temperature distribution seldom deviates from the inverse-distance law.

Of course, experts try to formulate disk models that fit reality best. In particular, a parameter for the disk viscosity is introduced in order to calculate how nebular matter that falls from interstellar space is absorbed, and at what rate the central star is fed. As complex as they may be, models still oversimplify the unknown. For example, because it is not known whether the accretion rate varied, models will assume that it remained constant. A simplified law is also adopted to describe the friction of the adjacent layers of gas. The essential point, for our purpose, is that these approximations have very little effect on the radial distribution of temperature in the median plane of the disk, at the time of dust settling out of the gas.

This theory can be partially verified by empirical deductions. For instance, the formation temperatures of the planets were deduced from their average density and their internal structure, which were inferred from their rotational momenta. From these data, the American astronomer John Lewis found that the formation temperatures vary as the inverse of the planetary distances to the Sun. Although this method confirms the previous results, it remains questionable, since it relies on the uncertain inner structure of the planets.

Fortunately, in contrast to the planets, the history of the chondrites was not complicated by episodes of thermal differentiation, so that it can serve as a 'cosmic thermometer' that was working at the exact time of the settling out of dust and was not perturbed by any later reheating. The working of this cosmic thermometer is based on the volatility of metals. Since all the non-volatile metals are present in chondrites with relative abundances identical to those of the Sun, it seems obvious that the losses of volatile metals such as lead, bismuth, tantalum and indium, compared with solar abundances, must have been caused by maximum heating of the dust during the epoch *before* it was enclosed in the chondrites, i.e. at the time when it settled out into the equatorial plane of the accretion disk. The observed losses are indeed in proportion to the volatility of the metals, so that the vaporization temperature for each of the volatile metals must have been the same. This confirms the reliability of our 'cosmic thermometer' at the time of the dust accretion.

From this 'fossil cosmic thermometer', the American meteorite specialists John Larimer and Ed Anders independently established the formation temperatures for different kinds of chondrite. They confirmed their results by several independent techniques that I do not need to describe here.

It is possible to identify the source of carbonaceous chondrites with the black C class asteroids (which become numerous beyond 2.6 AU), and the ordinary chondrites with the bright S class asteroids (which are more numerous at lesser distances). A surprising confirmation of these distances was given in September 1996 by the planetary scientist Mike Gaffey, who claimed that a single S class asteroid, the 185 km Hebe, is the sole source of fully half of the known ordinary chondrites. It happens that Hebe is at an average distance of 2.424 AU from the Sun, which puts its period of revolution sufficiently close to the 1 : 3 resonance with Jupiter's period to explain why so much of the debris of Hebe's collision with another asteroid was scattered by Jupiter's influence and eventually reached the Earth.

Coming back to our cosmic thermometer, we can conclude that, at the epoch of dust settling in the equatorial plane of the accretion disk, its temperature was about 450 K at a distance of 2.6 AU from the early Sun. This result was used in Figure 4.5 to standardize the temperature distribution in the mid-plane of the accretion disk, at the epoch when dust was imprisoned in the chondrites. The inverse law of temperature dependence on distance is then achieved by drawing a 45° line through the point of 450 K at 2.6 AU, made possible because the scales are logarithmic.

Figure 4.5 shows that the dust which formed the Earth was very hot. In the zone from 0.8 to 1.3 AU, it was between 900 and 1400 K. In that range of temperature, dust could not contain any residual water or volatile matter. The reason is that, during their sedimentation, the microscopic grains of anhydrous silicates or of metallic iron remained in close contact with a gas, at about 1000 K, consisting mostly of hydrogen. The frost formed from water and organic matter that covered the grains had evaporated a long time since, giving rise to superheated carbon monoxide and nitric oxide that had diffused through the scorching gas.

Between 2.6 and 5.2 AU, however, the temperature had remained low enough not to vaporize volatile organic substances. Finally, beyond 5.2

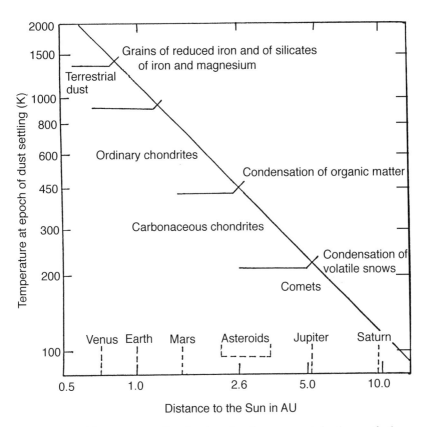

FIGURE 4.5 Temperature distribution in the equatorial plane of the accretion disk of the solar system, at the epoch when dust had settled out from the nebular gas. The distance 2.6 AU, which still separates the black C class asteroids from the light S class asteroids, corresponds to the temperature of 450 K that separates the formation of the carbonaceous chondrites (less than 450 K) from that of the ordinary chondrites (more than 450 K). This 450 K temperature and the corresponding distance of 2.6 AU have been used here to standardize the temperature distribution, whose dependence on distance at that particular epoch can be deduced from the virial theorem. Hence, it seems that the Earth was formed from hot grains of iron and silicates that were totally degassed as they settled. Water remains in the gas as steam, carbon is entirely in the form of carbon monoxide and nitrogen as gaseous N_2. At the same time, snows condensed on embryonic Jupiter.

AU, the water snow no longer evaporates. At below 225 K, it survives in its solid phase in empty space. It is not just coincidence that 5.2 AU is also the distance of the first giant planet. The presence of water ice weighed down the available solid matter and increased its mass enough to induce the emergence of a huge planetary embryo – the future Jupiter.

Formation of the planetesimals

The settling out of the dust is caused by sedimentation from the gas in the accretion disk, and this constitutes a fundamental branch point for the possible build-up of rocky planets, as well as an important step in the formation of the giant planets. After sedimentation of the dust, its density increased very much in the equatorial plane of the disk. A that time, it formed giant but extremely thin rings around the Sun.

Dust sedimentation from the gas is a gravitational phenomenon. As the dust grains revolve on quasi-circular orbits around the Sun, the solar force of attraction is restricted to maintaining them on their orbits. On the other hand, there is a much weaker force coming from the mass of the accretion disk; this force draws the grains toward its equatorial plane. Unimpeded, the resulting motion would be an oscillation across the central plane, but the viscosity of the gas damps it out rapidly. Hence the dust grains accumulate within the central plane.

Because of this accumulation of grains, the density becomes so large that the grains often collide, and tend to stick together; they aggregate first into fluffy balls of increasing sizes. This accumulation goes through many stages, which depend on both the degree of adherence of larger and larger aggregates, and also on the degree of residual turbulence in the gas.

If there were no turbulence left, objects of a few kilometers' diameter would form in a few years or scores of years, through the weak action of their own gravitation. But since there must be at least some residual turbulence in the gas, the process may be considerably slowed down, and it may take from 1000 to 100 000 years to form objects of more than 1 kilometer diameter. At any rate, this is still a very short period in the time scale of the Solar System.

We can assume that after 10 000 years, there are, everywhere in the central plane, irregular objects varying in size from 1 to 10 km. These objects are called '*planetesimals*'; they orbit the Sun in thin giant rings that look very much like the present rings of Saturn. At that time, they remained surrounded by a much thicker disk of nebular gas (mostly hydrogen and helium). The nature of the matter present in the planetesimals changes with their distance from the Sun, because they have been formed by particles whose nature was modified by the ambient temperature.

The zone in which the terrestrial planets formed was very hot. In the inner part of what was going to become the asteroid belt, the silicate particles not only lost their frost, but they were totally dehydrated and degassed, every volatile compound being lost. In the outer part of the asteroid belt, frost had also disappeared, but the interstellar grains were not heated enough to destroy the preexisting organic matter. Finally, at distances beyond 5.2 AU, the temperature never went beyond 225 K, and the surface frost of the grains never evaporated. Beyond 5.2 AU, the planetesimals therefore contained large amounts of water snow: these objects were cometary nuclei in their primitive stage. Revolving around the Sun in more or less circular orbits, these icy planetesimals remained pristine; only those whose orbits were eventually scattered and which penetrated the inner Solar System developed a tail and became ordinary comets.

Beyond the zone of the giant planets, from 30 AU to several hundred or 1000 AU, the icy planetesimals formed an immense ring of billions of objects outside the planetary system. This residue of the accretion disk formed a giant belt of icy planetesimals, which has hardly changed for more than 4 billion years. It is called the Kuiper belt, after the late American astronomer of Dutch origin, Gerald Kuiper, who predicted its existence. This belt is the source of all the short-period comets. These objects, detached from the belt by orbital resonances with the period of Neptune, enter and decay fast within the inner Solar System. Recently, more than a score of large icy planetesimals have been identified within the Kuiper belt (see Figure 4.6).

In contrast, the planetesimals that were within the zone of the future terrestrial planets were rocky (within 2.6 AU) or rocky-carbonaceous (from 2.6 to 5 AU). Their sizes were such that their gravity remained 1000 to 10 000 times weaker than our terrestrial gravity. This was enough to compact the dust slightly, but totally insufficient to capture

FIGURE 4.6   The nucleus of Halley's comet, photographed by the Giotto
space probe, at the time of its close encounter of 14 March 1986. The
nucleus, whose maximum length is 15 km, appears to have been formed
by the agglomeration of two planetesimals, each of 8 km diameter. Its
volatile snows are protected by a surface layer of silicates, which
presumably also contain carbon and organic matter because it is very
black. This crust must have been formed by the vaporization caused by
multiple passes at a short distance from the Sun, during the last
thousands of years. When Halley's comet nears the Sun, it still
continues to vaporize snows and to eject gas and dust, but only from a
set of craters that pierce the surface crust.

gas from the solar accretion disk. This basic separation of dust from gas is one of the most important clues provided by the study of chondritic meteorites.

The scenario for the origin of the planetary material has been confirmed not only by numerous chemical clues in the chondrites, but also by the discovery of many dust disks around very young stars still in formation (see Figure 4.3). The probable existence of these dust disks was suggested by the existence of an excess of infrared radiation around a large number of very young stars, observed by the Infrared Astronomical Satellite (IRAS).

This excess of radiation has a maximum intensity around 100 μm wavelength and the intensity distribution suggests thermal radiation from rather cold bodies with a very large area, for instance a dust ring. The radiating area of such a ring could be very large without implying a large mass, because grains can be very small.

We are now able to take pictures of some of these disks in visible light, while obscuring the central star, which is too bright. The first trial in 1984 revealed (see Figure 4.7) a most spectacular object: the dust disk surrounding the star Beta Pictoris. Among the nearby stars that also possess dust disks, Vega (in Lyra), Fomalhaut (in Piscis Austrinus) and Epsilon (in Eridani) must be mentioned. By 1993, more than a hundred disks had been found, all around very young stars, in particular around stars in the process of formation in Orion's giant molecular cloud, part of which is shown in Figure 4.3.

The dust disk of Beta Pictoris has a diameter of more than 800 AU around the central star. Since these dust disks seem to be a necessary stage before the formation of planets, they provide us for the first time with an idea of the large number of planetary systems in formation around stars like ours. This was an important step before the progressive discovery of a multiplicity of large planets around nearby stars, which only started in earnest in 1995–1996. In this respect, new developments in the discovery of other planetary systems are expected soon, in particular with NASA's Hubble Space Telescope, not only successfully refurbished in December 1993, but also equipped with a new infrared spectrograph in February 1997; in future, the two Keck telescopes installed on the Big Island of Hawaii will be able to work together as a single interferometer.

FIGURE 4.7   Photograph of the dust ring surrounding the star Beta
   Pictoris, first observed by Bradford Smith (University of Arizona) and
   Richard Terrile (Jet Propulsion Laboratory). The disk is seen almost
   exactly in cross-section. In order to perceive the disk in visible light, it
   was necessary to obscure the light from the central star, which was 1000
   times too bright. The existence of the disk had been detected by the
   IRAS satellite because of the large excess of its infrared radiation. In the
   far infrared, the star's light is 1000 times fainter, which explains why the
   thermal radiation of the cold dust dominates the signal received by
   IRAS. The diameter of the dust disk is 800 AU and displays quite well
   the features of the dust disk that preceded the emergence of planets in
   our Solar System (see also Figure 4.4.)

The T Tauri wind of the Sun

A hundred thousand years after separation of dust from gas, and dust
sedimentation into giant rings around the early Sun, this early Sun had
already reached its maximum luminosity. Despite a lower surface temperature (4000 K, rather than 5800 K for the present Sun), the early Sun

was four or five times brighter than at present, because it had an apparent surface area some 20 times larger than today. It was in the process of ending its contraction phase, so that it would reach what astronomers call the 'main sequence' in 20 to 30 million years.

A large number of the young stars of low mass recently discovered with dust rings – remnants of accretion disks – also show violent winds that escape from the stars. Although much more intense, this wind can be compared to the 'solar wind', which still escapes from the Sun. The first star discovered displaying such a wind was the T star of the constellation Taurus, called T Tauri. All stars showing this type of strong wind are now classified as T Tauri. The T Tauri wind is rather like a hurricane; in a few million years, it dissipates a huge amount (several percent) of the stellar mass.

The T Tauri wind seems to emerge at about the time when the accretion disk stops feeding the star. At the end of the accretion period, the star may rotate too fast, which threatens its own stability. The T Tauri wind which then appears is probably the mechanism that takes away the excess angular momentum and slows down the star's rotation; it must be an equatorial wind ejected by centrifugal force. The wind collides with the inner part of the accretion disk, and stops the growth process of the star.

The collision deflects the equatorial wind into two streams, to the north and to the south; presumably, the ions of the wind are bent by the star's magnetic field toward the two poles of the star. The shock wave of their encounter with the spreading accretion disk blows away, in a matter of 2 or 3 million years, the residual gas and the fine dust still present in the disk.

Only the preexisting planetesimals are not affected by the T Tauri wind, because their masses are already too large to feel this rarefied wind. At that stage, the average mass of the planetesimals is some 10 billion tons (10^{16} g). Moreover, the accumulation of planetesimals into larger and larger objects has already begun by the time the T Tauri wind starts to blow.

Agglomeration of the planets

At the beginning, the planetesimals were some hundreds of billions of irregular objects a few kilometers in diameter. The nucleus of Halley's comet, shown in Figure 4.6, is a good example of what a planetesimal looked like. After all, pristine planetesimals have been stored in the Kuiper belt for almost five billion years now, and many comets are their decaying remnants. Before being blown away by the T Tauri wind of the early Sun, the nebular gas surrounding the planetesimals was still 99% hydrogen and helium, with less than 1% carbon monoxide, some nitrogen and some hydrocyanic acid. This gas slowly cooled; let us keep in mind, however, that it could no longer interact chemically with the inside of the planetesimals.

The planetesimals were in quasi-circular orbits around the Sun, all in the same equatorial plane. Since they were very numerous, they withstood countless collisions; and since the orbits rotated in the same direction, collisions occurred at very low relative speeds and the objects did not bounce; they gently crashed into one another and agglomerated. It is a process not essentially different from the accretion of dust, except that the planetesimals began to have a weak gravitation.

Calculation shows that, after countless collisions, a smooth size distribution appeared, showing a scale invariance (a fractal again!). More specifically, the mass distribution is given by $dN = m^{-1.8} \, dm$, where N represents the number of objects of mass m. The exponent -1.8 remains the same for all sizes. The largest objects of such a distribution continued to grow by further agglomeration, and accumulated the largest fraction of the available mass.

Let us first consider the zone of the terrestrial planets (from Mercury to Mars and beyond). After 100 000 to a million years, this zone may have held about 20 objects that had already reached the size of the Moon. Of course there were still millions of smaller objects, of all intermediate sizes right down to the size of the planetesimals.

In its final stage, the nature of the accumulation changed. Because their attraction was no longer negligible, the gravitation of massive bodies changed the form of the trajectories. Two different effects appeared. First, massive objects attracted smaller objects from afar, inducing a larger number of collisions, hence a speeding up of the accumulation of larger objects. Secondly, those smaller objects which were deflected but had avoided collision during a grazing pass escaped along more elliptical orbits and shifted from their initial plane. This means that, by their next collision, the relative speed had increased, increasing also the chances of fragmentation as opposed to agglomeration.

Massive objects introduced therefore a branch point in the growth of the two populations: the growth of massive objects speeded up; the growth of smaller objects slowed down, and eventually stopped when fragmentation and accretion rates became equal.

The giant planets

An acceleration in the growth of massive objects occurred first at the point at which the sedimentation of solid particles led to a larger amount of icy grains, because the accretion disk was sufficiently cold, particularly where Jupiter was going to form.

This effect is not negligible, as demonstrated by the comparison of the amount of solids at different distances from the Sun, i.e. at different temperatures. Above 450 K, only the grains of silicates and of iron are solid. Their mass accounts for about 0.3% of the gas mass. Between 450 K and 225 K, organic compounds add weight to the grains, which then account for 0.5% of the gas mass. Beyond 5.2 AU, the temperature drops below 225 K; there, water-ice loads the grains so that the solid phase becomes 1.8% of the gas mass, which is a mass six times larger than in the zone of the terrestrial planets.

One would think that this effect would lead to a planet of about six terrestrial masses at 5.2 AU, where Jupiter is located. But there is another factor: volatile elements vaporized from grains located in the warmer zones inward of 5.2 AU and the vapors were transported outward by gaseous diffusion. These volatile vapors condensed on the 'cold wall' of the first icy grains they met, which were at 5.2 AU. Thus, it is not surprising that an enormous planetary embryo of some 12 terrestrial masses soon built up at the distance of the future Jupiter. Indeed, that planet appeared in less than a million years, thanks to the 'runaway' growth phase of the massive objects.

On the assumption that the gaseous phase of the accretion disk is still present, theory predicts that an embryonic planet need only be larger than 10 terrestrial masses to develop a gravitation capable of capturing an immense atmosphere very quickly. This explains the huge gaseous atmosphere of almost 300 terrestrial masses of hydrogen and helium that still envelops Jupiter.

Such a scenario has been confirmed by observational data. The American astronomer William B. Hubbard used the average density of

Table 4.2. *Masses of the giant planets*

The masses of the giant planets have been separated into the masses of their solid cores (assumed to be metallic silicates, iron, carbon and oxygen) and the masses of their envelopes (mostly hydrogen and helium). The solid cores all exceed the critical mass of 10 terrestrial masses. This 'critical' mass is the minimal mass required to capture a gaseous atmosphere from the nebular gas in the accretion disk, through gravitational attraction by the embryonic planet. The amount of gaseous atmosphere of the outer planets subsides rapidly with their growing distance from the Sun. This comes from the fast dissipation of the nebular gas during the accumulation of the solid cores. These embryos were accreted more and more slowly, in proportion to their lengthening periods of revolution, as one moves out through the Solar System.

Planet	Core (M)	Envelope (M)	Total mass (M)
Jupiter	29	288	317
Saturn	19	75	94
Uranus	13	1.6	14.6
Neptune	15.4	1.9	17.3

Note:
The mass M of the Earth is used as unity.

Jupiter and its flattening by rotation, to show that this planet still contains a solid core, denser than that which the compression of its atmospheric gases alone should yield. If this core is made of silicates, iron, carbon and oxygen, it must amount to 29 terrestrial masses.

After 2 million years, this scenario occurred again, with the runaway growth of a solid core for Saturn; but, in that epoch, the gaseous fraction of the accretion disk had already been partially dispersed by the action of the T Tauri wind of the Sun. This is why Saturn could only accumulate an atmosphere of 80 terrestrial masses, around a solid core of the same magnitude as that of Jupiter.

Finally, the agglomerations of Uranus and Neptune were even slower, because the durations were in proportion to their periods of revolution around the Sun. The embryos of the two giant outer planets became sufficiently massive only after all the gas from the accretion disk had been dissipated, so that it was too late for them to accumulate a substantial atmosphere.

Thus, the origin of the giant planets has been quite well elucidated, and the particulars of their internal structure have been explained by their history. The formation of solid planetary embryos was caused by the accretion of snowy planetesimals which were nothing less than cometary nuclei. The mass of these embryos exceeded the critical amount needed to induce the rapid capture, by gravitational attraction, of huge amounts of nebular gas; but the T Tauri wind quickly dissipated this nebular gas. Since the epoch when the critical size of an embryo is reached depends on the periods of revolution, the mass of the atmosphere of a giant planet shows the relative depletion of the available gas, at the time when the planet's embryo reached its critical mass (Table 4.2).

Asteroids and comets

There is evidence in the asteroid belt that the slower growth regime was set up early on, and that it checked and eventually stopped the formation of normal planets. What happened? In order to understand this, let us return to the moment when the embryonic Jupiter reached 10 or 12 terrestrial masses. This was carried out by a runaway accretion. Beyond this, Jupiter's mass grew steadily to more than 300 terrestrial masses through capture of its atmosphere.

At that time, billions of icy objects still remained in circular orbits around the Sun in the zone of Jupiter. These objects were indistinguishable from comets, but their size distribution was changed by a steady accretion; most of their mass was already in objects of 50 to 500 km diameter, although billions of smaller comets were still present.

Soon, Jupiter displayed a larger and larger mass. Not only did the planet continue to absorb comets (which explains why its solid core finally reached a mass of nearly 30 terrestrial masses), but it perturbed more and more the orbits of those that it did not manage to capture. The 'escape velocity' from Jupiter's gravitation is 61 km/second; this is also a measure of the extent by which Jupiter can modify the velocity of a grazing orbit. This means that, as Jupiter's mass grew steadily to a very large value, it became able to eject comets in all directions, at speeds that considerably exceeded the escape velocity from the Solar System itself, which at Jupiter's distance is 18.5 km/second. Thus began the long diffusion process of those cometary orbits, which were originally almost circular and happened to be in Jupiter's zone. This process was repeated at slower and slower rates for the outer giant planets.

This type of process has several effects:

(1) In the different zones of the giant planets, most comets were eventually thrown out of the Solar System and lost forever. This ejection took place at random, in all directions, in particular through the inner part of the Solar System.

(2) However, a fraction of these comets was not lost; they stayed in an enormous sphere of some 50 000 AU radius, surrounding the Solar System. These comets remain held by the Sun's gravitation because other possible influences by nearby stars are negligible. They constitute what is called the 'Oort cloud', after the Dutch astronomer Jan Oort, who established its existence. All very long period comets observed today came originally from the Oort cloud, from where they were perturbed long ago, mostly by galactic tides.

(3) During their ejection from the Solar System, a large number of comets crossed the asteroid belt, where they perturbed asteroid orbits. As a result, these orbits became more elliptical and more slanted. Further collisions among asteroids occurred at larger relative speeds, and resulted in less accretion and more fracturing and rebounding. For this reason, the growth of the asteroids' size stopped, ruling out the emergence of an Earth-size planet in the asteroid belt. This also limited the growth of planet Mars to barely more than 10% of the mass of the Earth.

(4) Since the comets were ejected in all directions, a good proportion of them escaped by first crossing the inner solar system, bombarding all objects, planets and satellites that were in their path. The traces of this bombardment are still visible as impact craters on every object with a solid surface, namely the Moon, Mercury, Venus, Mars and its satellites, to which we can add the satellites of Jupiter and Saturn (the lunar cratering is illustrated in Figure 5.1).

The terrestrial planets

During the dissipation of the nebular gas, several scores of planetary objects had time to form in the zone of the terrestrial planets (Table 4.3). These 'protoplanets' were too big to be called planetesimals any longer; they had masses of from 1% to 10% of the mass of the future Earth. There remained also millions of planetesimals of all sizes, from, say, 5 to 500 km.

The most massive objects reached the stage of a runaway agglomeration because of their strong gravitational attraction (they are from lunar size to Mars size). The distant presence of Jupiter insidiously perturbed the eccentricities of all these objects, which means that they could not stabilize on a few score of circular orbits. Sooner or later, they collided again, but these giant collisions diminished their number and consolidated a few of them, until the four terrestrial planets stabilized in quasi-circular orbits that still felt Jupiter's presence, but would never intersect again.

The epochs of the last large collisions are still quite uncertain because they resulted from a random process with a very small number of objects. At that time, it is possible that some moderate-sized asteroids were sufficiently perturbed to bring volatile matter, e.g. carbonaceous chondritic material, to the Earth. The heat of the final collisions could possibly have degassed the Earth again. This is also the time when there could have been two or three objects of the size of Mars travelling along unstable orbits, constantly perturbed by close encounters with neighboring planets. After 40 million years, the Earth was probably close to its final mass. During the 100 million years that followed, in addition to countless small impacts, there were at least two giant collisions: one with Mercury, the other with the Earth. The latter led to the formation of the Moon. Let us examine the two cases by going back in time.

Agglomeration of the Earth

Around the largest objects, a growing gravity raised the relative speeds of impact, hence the final stages of the Earth's agglomeration involved greater and greater speeds of collision. The speed grew in proportion to the radius of the Earth, and the heat released by each impact

Table 4.3. *A possible chronology for the solar system*

This chronology is approximate, but compatible with all known data. It represents well the present state of knowledge. For instance, the dissipation of the nebular gas is possible in 3 to 6 million years, because within such a time scale the solar T Tauri wind can dissipate a mass of 5% to 10% of the mass of the Sun, which is comparable to that of the nebular gas. The growth of the extended atmosphere of Jupiter and Saturn was possible because it occurred before the end of the dissipation of the nebular gas, whereas the growth of the atmosphere of Uranus and Neptune was insignificant because it occurred too late (see Table 4.2). The half-life for the diffusion of cometary orbits is the period during which the number of still extant orbits fall by half. This is an exponential decrease, meaning that $\frac{1}{4}$ of the orbits survive after two half-lives, $\frac{1}{8}$ after three half-lives, etc.

Dust sedimentation from gaseous disk (chondrite ages set it at 4.56 billion years ago)	Age zero
Planetesimals reach 10 km	10 000 years
Planetesimals, size distribution 50 to 500 km	100 000 years
Runaway agglomeration for Jupiter's embryo (10 Earth masses)	1 million years
Thirty-odd protoplanets from Moon- to Mars-size in the zone of the terrestrial planets	1 million years
Runaway agglomeration for Saturn embryo	2 million years
End of dissipation of nebular gas	5 million years
Runaway agglomeration for Uranus embryo	7 million years
Runaway agglomeration for Neptune embryo	14 million years
Accumulation of the Earth 99% finished	40 million years
Formation of the Moon	50 million years
Half-lives for orbital diffusion	
Comets from Jupiter's zone	50 million years
Comets from Saturn's zone	120 million years
Comets from Uranus's zone	300 million years
Comets from Neptune's zone	600 million years
Cometary bombardment of the Earth is 99% complete	1 billion years

increased with the square of this speed. Hence the heat released on impact grew in proportion to the square of the Earth's growing radius.

Imagine an ideal case where a planet agglomerates from collisions with small homogeneous objects: the center of the planet is cold, while the temperature of the outer layers is becoming higher and higher, rising in proportion to the square of their distances from the center. Eventually, the center will be heated by conduction from the outer layers, but conduction is a slow phenomenon. Melting temperature is reached first at the surface of the planet, and liquid lava appears at the impact points and eventually covers the surface area of the whole planet.

The Earth had practically reached its final mass when the high temperature of the outer layers managed to propagate sufficiently toward the center to liquefy the particles of metallic iron that were mixed with silicate grains. Because iron is denser than silicates, molten iron started separating from the lighter silicate, and seeping throughout the Earth's mass to its center. As in a blast furnace, the light silicate slag floated up, while the melting iron flowed down. Seismography can still detect the central iron core which has differentiated from the silicate mantle.

In the past, geologists argued about the need for a reducing agent like carbon to reduce iron to the metallic state before the separation could take place. This is pushing the analogy with the blast furnace a step too far. We have now discovered that there were grains of metallic iron in interstellar space. These iron grains were already in the reduced state when they collapsed with the nebular gas onto the accretion disk. When the dust separated from the gas, they were immersed in a large excess of hot hydrogen, a better reducing agent than carbon, at about 1000 K in the Earth's accretion zone.

As far as planet Mars is concerned, the lower temperature of the grains at the time of accretion suggests that the central core could be formed from a eutectic mixture of iron and iron sulfide, although this remains uncertain.

The presence of radioactive elements, especially uranium, made in supernovas and brought to the inside of the Earth by interstellar grains, did not have much influence at the outset, since radioactive nuclei like uranium require eons for their heat to be freed. But after long geological periods, this contribution to the inner heat of the Earth became more and more important. There is not much uranium in total, but its heat production is very large. Think of the atom bomb, where nuclear heat is released immensely fast! Uranium turned out to be essential for explaining the present thermal balance of the Earth.

The epoch of formation of the metallic central core of the Earth remains rather uncertain, but its make-up can be assumed to have taken place progressively between 40 and 60 million years after the age zero defined by the settling out of the dust from the nebular gas. The reason for allocating such a late epoch is that the heating was not large enough to melt iron before the Earth had almost reached its final mass.

Formation of the Moon

The formation of the Moon has remained an enigma until recently. The Moon's chemistry and geology seemed difficult to explain. Its density of 3.3 implies that it does not possess a sizeable iron core, but the missing iron is not found anywhere; notably, it is not incorporated into the silicate mantle. In the lunar stones brought back by astronauts, there is also a puzzling deficiency in the usual volatile metals (sodium, potassium), compounded by a perplexingly large concentration of refractory oxides (those of aluminum and calcium). It looks as if terrestrial rocks had been brought to high temperature, then placed on the Moon; otherwise, they appear very similar to the silicates on the Earth.

The collision with the Earth of a protoplanet just a little heavier than planet Mars could have detached the Moon from the Earth. This was proposed by the American astronomers Alistair Cameron and William Ward in 1976. However it seemed an unlikely random event, until it was studied in great detail numerically, by calculating at each moment the distribution of masses and angular momenta, as well as the fate of the different fragments in their trajectories.

To be effective, the collision would have to have been very slow; this implies a protoplanet in quasi-circular orbit, in the immediate vicinity of the Earth's orbit. The collision had to occur at a near grazing angle, after the iron core of the Earth had already been separated. These two conditions are required to explain the Moon's deficiency in iron and the angular momentum needed to stabilize the Moon's orbit.

The collision would have detached from the terrestrial mantle a fraction of silicates of the order of 10 times the lunar mass. This mass would have been brought to high temperature by the collision before being dispersed into space. A single lunar mass would have remained in orbit around the Earth, in the form of multiple irregular fragments spread into a close ring, comparable on a smaller scale to the rings of Saturn. The fragments of this transient ring would have agglomerated progressively into a single object. At first the orbit would have been very close to the Earth, but it would then continue to broaden the distance up to the present day. It has long been known that the Moon is progressively slowed down along its orbit by the terrestrial tides, and that it continues to spiral away from the Earth. At the present time, laser measurements of the Moon's distance show that it continues to move away at the rate of 4 cm per year.

It is difficult to establish the age of the Moon with accuracy. The collision occurred after the differentiation that formed the Earth's iron core, but it could have occurred much later. A protoplanet of the size of Mars could have survived on an unstable orbit for 100 million or 200 million years. The American planetary scientist George Wetherill remarked in 1975 that a clue is given by the oldest lunar rocks. The lead isotopes derived from uranium decay point to a final solidification of the oldest lunar rocks at about 130 million years after age zero. If we allow 10 million years for the Moon to agglomerate and solidify, this would suggest that the collision that formed the Moon took place 120 million years after age zero. However, data derived from lunar rocks in November 1997 suggest that the Moon may have formed 50–70 million years earlier; this is not inconsistent with the other data given in Table 4.3.

Fate of the planetesimals

One hundred million years after age zero, there were still a large number of planetesimals not agglomerated into planets. However, sooner or later their fate would be to hit one of the existing planets. That is if they did not escape from the Solar System and finish up in outer space, after their orbit was deflected by a planet of sufficient mass.

After a few billions of years, the rare surviving objects were parked in stable, quasi-circular orbits that prevented close encounters with massive objects. Small pristine planetesimals were unlikely to have survived in significant numbers, even in the asteroid belt, where a very large number of collisions may have fragmented them.

As for the icy planetesimals that we call comets, there are still some on quasi-stable orbits between the giant planets, where a few objects of moderate size have recently been discovered on almost circular orbits. Finally, since 1992, a few scores of rather large objects have been discovered beyond the orbit of Neptune. They are believed to be the largest of an extremely large number of icy objects. In the section on the formation of the planetesimals (see p. 91), I have already mentioned that such a belt of comets was predicted by Gerald Kuiper. His prediction was based on the mass distribution of the planets in the Solar System. There was a 'missing mass' beyond Neptune. It has been demonstrated recently that this Kuiper belt is also the source of the 130 short-period comets observed during recent centuries. All short-period comets have been perturbed out of the Kuiper belt and into visibility from Earth by resonances with the period of revolution of Neptune. Their existence implies that millions of undetected cometary objects still survive in the Kuiper belt.

On the other hand, there are probably about a hundred billion comets in the Oort cloud, where they are parked on stable orbits and are not observable from Earth because of their great distances. Each year, however, one or two new comets appear that have never been observed within the planetary system. They are on very long-period orbits, and their trajectories show that they come straight from the Oort cloud, from which they have been detached, either by a neighboring star passing through the cloud, or more often by the 'galactic tides' that perturb its stability.

The Oort cloud is a quasi-spherical reservoir of comets centered on the Sun and 1000 times larger than the widest planetary orbits. It stores the residue of those comets expelled by the giant planets during their growth. The formation temperatures of these comets are given in Figure 4.5: 110 K for Saturn's zone and 220 K for Jupiter's zone. Since their snows condensed from the solar nebula at these temperatures, the isotopic exchange of H_2O (or HDO) with the nebular H_2 (or HD) produced a deuterium (D) enrichment that varied with temperature (8 times at 200 K, 60 times at 100 K). The Earth's oceans also have a deuterium enrichment 8 times that of the nebular gas (which is known because it is still extant in Jupiter's atmosphere); this suggests that the oceans originated from comets coming mostly from Jupiter's zone, with a minor fraction from that of Saturn.

In contrast, the Kuiper belt comets, source of the present short-period comets, must be more pristine, since they kept their snows at 30 to 40 K, mostly as they were in interstellar space. Their larger deuterium enrichment will depend on their location in the belt, and on their possible isotopic exchanges.

The planetary system

Three hundred million years after age zero, the planetary system was well established. It included the four giant planets, the four terrestrial planets, a number of smaller objects like the Moon, the many satellites and innumerable asteroids. Comets were in the Kuiper belt, or slowly being stored in the Oort cloud, since their orbital diffusion from the zones of the giant planets was not yet finished. The planet Pluto might be a huge icy planetesimal which, before stabilizing on its present orbit, had an early passage close to Neptune. During this passage it could have captured one of Neptune's satellites, and this could explain the origin of Pluto's present satellite.

The satellite systems around giant planets probably originated from minor accretion disks, and the extant rings around Saturn, Uranus and Jupiter could be their residues. In capturing the nebular gas, Saturn and especially Jupiter heated it sufficiently to make it shine for a while like a modest star. Although transient, this early radiation was large enough to deplete volatile compounds on the moons – totally on Io and partially on Europa, although Ganymede and Callisto were too far away and retained their ice.

The terrestrial planets are properly called the 'rocky' planets, because they are made mostly of silicate rocks, covering a central core consisting mainly of metallic iron. As mentioned before, this core separation came from the existence of two types of grains, silicates derived from the oxidation of metals, and metallic iron grains derived from a reducing process elsewhere in interstellar space. It was the non-volatile fraction of these grains that survived the heat of the solar accretion disk.

Among the terrestrial planets, Venus and the Earth are by far the most similar in mass. However, their evolution differed from the beginning. Venus was not subjected to any large final impact, since it kept a much thicker atmosphere. The last two terrestrial planets are both much smaller. The mass of Mars is barely more than 10% of that of the Earth, and Mercury is less than half the mass of Mars.

We believe we understand quite well the process that kept the mass of Mars so small. The rapid appearance of a large protoplanet at the distance of Jupiter not only prevented the formation of a planet in the asteroid zone, but also prevented the accretion of a large number of protoplanets on Mars by scattering their orbits. This scattering of orbits may also explain the origin of the body whose collision with the Earth formed the Moon.

The abnormally high density of Mercury must have come from an impact, the trace of which has been found. This impact would not have been grazing, as happened to the Earth, but well centered, so that the iron nucleus of the impacting protoplanet would have penetrated deeply, eventually adding iron to the core of Mercury, whereas silicates from the mantles of the two objects would have scattered in space, diminishing the silicate-to-iron ratio (Figure 4.8, Table 4.4). On 29 March 1974, NASA's space probe Mariner 10 took a picture of the largest impact crater on Mercury, called the Caloris Basin. It has a diameter of 1300 km and is surrounded by a natural circle of mountains reaching an elevation of 2000 m. It shows several concentric circular rings, and must have been one of the latest impacts on Mercury by a particularly large protoplanet.

It is clear that the processes of final consolidation of the protoplanets into planets, whose final orbits are stable, depend too much on stochastic collisions for us to piece the facts together completely. Other planetary systems, while evolving in a similar manner, could assemble planets of different masses on different orbits. A typical example is the early existence of Jupiter, which increased the eccentricities of the orbits of a dozen terrestrial protoplanets. The broadening of the zones they swept out increased the masses of Venus and the Earth and created the four terrestrial planets instead of some ten smaller objects.

The eventual presence of planets like Jupiter or Saturn in other planetary systems is not an automatic consequence of the formation of an accretion disk around a star. The capture of a large atmosphere arising from the nebular gas depends on the early formation of an 'embryo' of more than 10 terrestrial masses *before* there is total dissipation of the nebular gas by the action of the T Tauri wind of the young star.

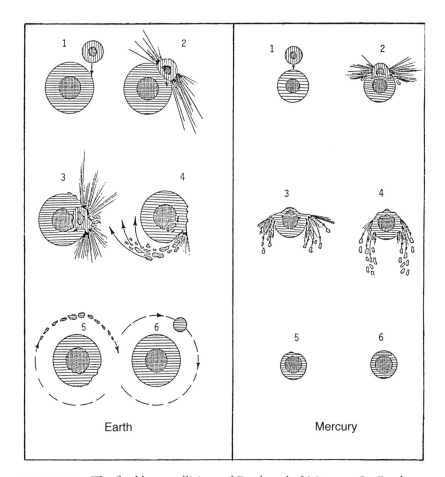

FIGURE 4.8   The final large collisions of Earth and of Mercury. On Earth, a grazing impact with a protoplanet the size of Mars penetrated the mantle and projected a portion of it into space. Only a fraction of this broken piece was captured in terrestrial orbit and ultimately formed the Moon. On Mercury, the impact was centered and the collision was faster. The fraction of the mantle ejected was completely lost to space; the central metallic core grew whereas the silicate mantle became smaller.

Table 4.4. *Percentage masses of the terrestrial planets*

The cores of the terrestrial planets are mostly metallic iron, with the exception of Mars, whose core could be a mixture of iron and iron sulfide. The mantles are composed mostly of silicates of iron and magnesium with small amounts of silicates of other metals; the mantle of Mars might be enriched in some iron oxide. The differentiation between cores and mantles comes from the impact heats that grow steadily with the growing mass, and hence the growing gravity, of the planet. Liquid iron flows to the center, whereas the much less dense silicates float to the surface of the iron melt. The lack of a large amount of iron at the center of the Moon, and the abnormally large amount of metallic iron at the center of Mercury, are explained by the circumstances of the final large impacts summarized in Figure 4.8.

Planet	Core	Mantle	Total
Mercury	3.2	2.1	5.3
Venus	28.5	43.0	71.5
Earth	38.0	62.0	100.0
Moon	0	1.2	1.2
Mars	2.8	7.9	10.7

Note:
The mass of the Earth is taken to be 100%.

However, the recent discovery of numerous giant planets very close to their central star (although a partial artifact coming from observation techniques which cannot yet detect smaller planets) could be an indication that we still do not understand the details of the processes that make planetary systems. A steady discovery of new planetary systems is expected in the near future and will shed new light on this problem.

The random variations in the gravitational collapse of an interstellar nodule that forms an accretion disk introduces stochastic differences in the distribution of masses and angular momenta. The infall rate may vary considerably, introducing temperature variations in the disk. These variables can lead to a wide variety of planetary configurations, including the place of formation of rocky or snowy planets, as well as planets of more than 10 terrestrial masses accreted too late to capture a giant atmosphere because of the dissipation of the nebular gas.

The possible absence of planets like Jupiter or Saturn may have other crucial consequences for the evolution of planets like the Earth. In particular, the diffusion of cometary orbits would extend to much longer periods. The atmospheres of the terrestrial planets which depend on the cometary bombardment induced by this orbital diffusion would not be completed within the first billion years. Continuing at a reduced rate for a much longer time, it is conceivable that this bombardment could perturb the conditions needed for the emergence of life.

These ideas are speculative and useful only to remind us that there are many unknown factors that could limit the emergence of life in many other planetary systems. In spite of the recent discoveries of several planetary systems around other stars, we must also keep in mind that we have not yet discovered any small rocky planet around a star of low mass of our solar type or approximating to it. Our present techniques are insufficient for the detection of objects so far away, of such low mass, so little and so dim as the terrestrial planets. But the situation is changing rapidly, and it is to be expected that more planetary-sized objects will soon be discovered around nearby stars.

5

THE EMERGENCE OF LIFE

The emergence of life is an expected phase transition from
a collection of polymers which do not reproduce
themselves, to a slightly more complex collection of
polymers, which do jointly catalyze their
own reproduction.

Stuart A. Kauffman, *The Origins of Order*, 1993

Evolution toward a growing complexity and organization
is the result of fluctuations that lead to a gradual
acquisition of autonomy from the environment.

G. Nicolis and I. Prigogine, *Self-Organization in Non-equilibrium Systems*, 1977

The origin of the biosphere

The biosphere is the ensemble of the life-supporting regions of the terrestrial globe. It is made up of the oceans and fresh waters, plus the atmosphere and the layer of soil (spread over the continents) that contains organic matter. Water is the predominant component of the biosphere, in which the atmosphere plays an important role, and where the many compounds of carbon are essential because they are needed by life. In the outer crust of the Earth, however, inorganic carbon is by far the most abundant component, in the form of carbonates (limestone, dolomite, etc.). Heat easily decomposes carbonates and frees carbon in the form of carbon dioxide (CO_2); thus this process is the principal source of volcanic CO_2.

For a long time, the origin of the biosphere remained a mystery, because no fossil trace exists from the first billion years of the Earth's evolution. Any memory of that period has been irretrievably lost for the geologists, so that any evidence of the processes that brought water into the oceans and supplied the atmospheric gases and the organic and inorganic carbon compounds has been completely obliterated.

The hypothesis that water, air and carbon compounds were *originally* produced through volcanism was first proposed by the Swedish geologist G. Hogbom in 1894. This century-old hypothesis has been accepted by geologists as the simplest explanation. Volcanic gases, although extremely variable, contain on average 60% steam and 30% CO_2; the remaining 10% is a mixture of sulfur dioxide (SO_2), H_2 and N_2. The absence of oxygen is not surprising, since oxygen is produced only by vegetation. Free oxygen would disappear fast if it were not constantly renewed by plants.

Volcanic gases are also constantly recycled by the existence of plate tectonics; when two continental plates collide, one is forced downwards and crushed. This triggers volcanic activity that heats water and dissociates carbonates. This explains the origin of steam and CO_2; sulfur comes from the sulfates dissolved in water. There is, however, a small fraction from 'juvenile' gases that might have been trapped in the Earth's mantle for a very long time. The amount of noble gases contained in this fraction may clarify their origin, because noble gases do not react chemically and so their history is easier to reconstruct.

The noble gases present in the deep-sea basalts, in particular helium and neon, suggest that the Earth's embryo captured first a 'primary' atmosphere, which was lost later. A 'primary' atmosphere is an atmosphere that would have been captured by the Earth's gravitation, if the Earth had acquired a sufficient mass *before* the nebular gas had been wholly dissipated by the T Tauri wind of the early Sun. Coming from the nebular disk, this 'primary' atmosphere would have been the same as that of the Sun, and this is this *solar* component that can be recognized in noble gases from the mantle.

Unfortunately, the large number of factors that govern the abundance distribution of the noble gases present in the deep-sea basalts leave large ambiguities. All that can be said is that, if this primary atmosphere did really exist, it was lost early on, presumably because the Earth's gravity was not yet strong enough to keep it when the T Tauri solar wind started to blow it away. We can conclude that some juvenile gases can still exist deep in the mantle, dating back to the time when a primary atmosphere left some noble gases in solution in the magma that was to become the mantle. However, their presence cannot explain the present volcanic activity.

Conversely, the distribution of the elemental and isotopic abundances of the noble gases still present in the atmosphere is easy to understand, and its interpretation is surprising: there is no trace left of a 'primary' atmosphere.

The composition of a primary atmosphere would initially have had to have been the same as that of the Sun, but, through the ages, the lightest gases would have been able to escape. However, that progressive loss slows down as the atomic or molecular mass rises; the decrease is regular, and the heavier gases are the least affected.

Among the noble gases, atoms of krypton and xenon are the most significant, because they are very heavy, and each gas has many isotopes. It was discovered that the six isotopes of krypton are present in regular proportions, and the seven isotopes of xenon are also in very regular, but entirely different, proportions in the atmosphere. The masses of the isotopes of krypton run from 78 to 86, and those of xenon from 124 to 136. Whatever the reason for these ratios, the observed abundances are totally incompatible with those in the Sun, and hence with the existence of a primitive atmosphere. These noble gases had to be carried to Earth; before their arrival in our atmosphere, therefore, their sources had to be separate, and independent of the gases in the primitive solar nebula.

The only known proportions that correspond to these conditions would be those of rare gases imprisoned in clathrates. Clathrates are hydrates of water ice which crystallize in a form different from that of ordinary ice with gaps in the crystals that can be filled by large atoms such as those of krypton and xenon. In collaboration with the Belgian astrophysicist Pol Swings, I proposed in 1952 that clathrates could explain the slow vaporization of gases from the cometary nucleus.

The properties of clathrates have recently been investigated by the Israeli physicist Akiva Bar-Nun; he has shown that the terrestrial isotopic proportions of the atmospheric noble gases correspond exactly to the low-temperature proportions due to the presence of ice clathrates at low temperature in comets.

Cometary bombardment

Geophysicists have never been able to explain the bizarre aportioning of isotopes of the noble gases present in our atmosphere in terms of a terrestrial origin. The primeval source of water in our oceans and in our atmosphere has been attributed to an early outgassing from the interior of the Earth, through volcanoes. This hypothesis does not rest on any observation. It is based on a blind extrapolation of the present processes, extended over the 'lost interval' of the first billion years (which is hidden from geology) of the formation of the Earth and its primitive history. In order to understand what happened during that lost period, we must leave geophysics and examine the astronomical evidence.

We need look no further than the Moon (Figure 5.1). The bombardment which explains the abundance of the impact craters that riddle the surface of the Moon has been dated; the period of greatest bombardment covers precisely the 'lost' interval on the Earth, and explains its cause. The bombarding flux that affected Earth is directly comparable to that of the Moon, because the Earth and the Moon are extremely close compared to the distances to the giant planets. The larger gravity of the Earth introduced a small correction factor, which is easily taken into account.

The cause of the lost interval in geology becomes obvious: all geological traces have been wiped out on Earth by the intensity of the bombardment. In the distant past, it was dominated by comets arriving from as far away as the giant planets (the diffusion half-time of their orbits is given in Table 4.3). It diminished by a factor of 1000 after the first billion years, and then asteroids became more prevalent. Thus, most of the water present in our oceans and our atmosphere was brought to Earth during the first billion years. Lunar craters provide fossil proof of the cometary origin of our biosphere. On the Moon, gravity was inadequate to hold any atmosphere for a long time; water vaporized and was lost to space, except in the rare places where ice has been perpetually protected from sunlight. In December 1996, radar detected packed ice hidden deep in craters near the south pole of the Moon; confirmed in early 1998 for both lunar poles, this is further proof of prior intense cometary bombardment.

The final phase of this bombardment has continued at a slower and slower rate up to the present time. In particular, all the planes of the orbits of the 130 short-period comets of the last three centuries, are rather close to the plane of the ecliptic. The plane of the ecliptic, which is the plane of the Earth's orbit around the Sun, is close to the average plane of the orbits of all the planets. The existence of the same plane of symmetry for the cometary orbits implies that the source of the short-period comets must be the Kuiper belt (which also has the same plane of symmetry) and cannot be connected to the Oort cloud (which is spherical). The process by which the short-period comets are detached from the Kuiper belt is also understood: their orbital periods are in resonance with Neptune's. The short-period comets correspond to the end of the diffusion of cometary orbits by Neptune, whose half-time of orbital diffusion is the slowest (Table 4.3). We know that all short-period comets are on unstable orbits because sooner or later they will end by cutting across planetary orbits. If not ejected, they will eventually collide with one of the planets. Collisions with the Earth should occur now with a frequency that varies from 10 million years for the numerous smallest comets to perhaps 1 billion years for the largest objects.

The other planets are not immune. The collision of fragments from comet Shoemaker-Levy 9 with Jupiter occurred as predicted in the week of 16 to 22 July 1994. The comet had been broken into a dozen fragments by its first grazing encounter with Jupiter, so that the fragments were dispersed along the orbit. The trajectory ended with a plunge into the atmosphere of the giant planet, and therefore contributed to the enlargment of Jupiter's central solid nucleus.

As regards our biosphere, the accretion model that explains the lunar cratering so well (see Figure 5.2) also predicts the amount of 'siderophile' metals present in the terrestrial crust. The siderophile metals are those with an affinity for metallic iron. If they had been present on Earth from the beginning, they would have blended in with the molten drops of iron and would have gone down with them to the Earth's core. Their presence at the surface of the Earth, and in particular their abundance ratios which have remained at cosmic levels, imply that they have been brought to Earth *after* the separation of the iron core. So we now assume that they were brought later by the cometary bombardment.

The model shown in Figure 5.2 also predicts the total amount of carbon present on Earth, including all that is present in the organic

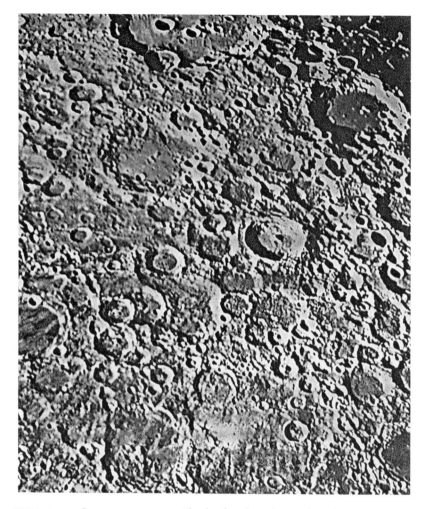

FIGURE 5.1   Lunar craters provide the fossil evidence that the terrestrial
biosphere (oceans, atmosphere and carbon compounds) has arisen
from a bombardment, 99.9% of which occurred within the first billion
years (see Figure 5.2). The long duration of the bombardment was a
consequence of the long periods of revolution of the giant planets,
which brought long decay times in the orbital diffusion of those
planetesimals that were present in the distant, cold zones of the Solar
System. In those regions, the planetesimals held large quantities of ices
of water and of other volatile compounds. When they came to the inner
Solar System, they developed a tail and we call them 'comets'. Lunar

compounds and also in the large amount of carbonates present in sedimentary rocks. Finally, the comets have brought to Earth 10 times more water than the present mass of the oceans, and 1000 times more gas than is now present in our atmosphere.

These last two values may seem surprising, but they are welcome because they are needed to explain the considerable losses of volatile matter resulting from the many giant impacts predicted by the theory of planetary accretion. In particular, the proposed grazing collision with a protoplanet the size of Mars, which would have formed the Moon, must have produced an enormous amount of heat, with complete loss of the water and the gas already amassed over the first 120 million years. We mentioned that there were probably a score of protoplanets larger than the Moon that remained for a long time on unstable orbits in the zone of the terrestrial planets; many of them were progressively ejected from the Solar System by grazing encounters, but it is reasonable to assume that there were also some collisions with other large objects.

Moreover, the many collisions with comets of all sizes (typically 5 to 500 km in diameter, judging from the lunar craters) imply a deep 'plowing' of the crust and of the upper mantle of the Earth. These collisions explain the origin and history of the volatile matter buried deeply in the ground, which led to the appearance of volcanic activity.

On the Earth, as on the Moon, the largest of these impacts caused a temperature increase sufficient to melt the surface silicates. This triggered the transient appearance of 'seas of lava' that could sometimes cover hundreds of thousands of square kilometers.

On the Moon, these seas of lava, whose locations are still quite visible (they are called 'maria', the latin name for 'seas'), locally reset to zero the radioactive 'clock' for the rocks that allowed us to establish the ages in Figure 5.2. On Earth, this deep plowing destroyed the fossil evidence of what occurred during the first billion years, to the great

Figure 5.1 (cont.)
gravity was not strong enough to retain any trace of that water and volatile matter coming from the cometary bombardment which brought our oceans and our atmosphere. The only regions where water-ice could have survived until now on the Moon is near the poles, in places perpetually protected from sunlight, where some was probably found in December 1996.

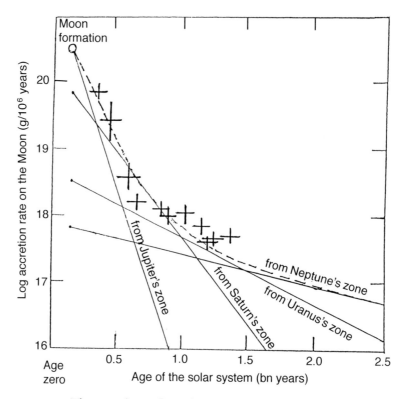

FIGURE 5.2   The cumulative flux of lunar craters has been computed for different regions of the Moon from the observed crater density. The age of each region is deduced from the solidification age of those lunar rocks that have been brought back to Earth (previous craters were erased by molten lava). The data come from the Basaltic Volcanism Study Project (1981) and the uncertainties are shown by the size of the crosses. The dashed line is the sum of the four exponential decay times for the number of comets coming from the four giant planets. The total mass of comets coming from each giant-planet's zone is derived from Safronov's (1973) analysis.

loss of geologists. Likewise, the greater intensity of the very early bombardment of the Moon prevents us from identifying lunar regions more ancient than those mentioned in Figure 5.2.

   Finally, there is more telltale evidence suggesting that terrestrial water was indeed brought to the Earth by comets. It is the isotopic ratio of deuterium to hydrogen in seawater – close to eight times the nebular ratio still preserved in Jupiter's atmosphere.

We noticed earlier that such an eight-fold enrichment occurs by isotopic exchange of deuterium between H_2O/HDO and H_2/HD gas at 200 K. This temperature is predicted in the accretion disk (see Figure 4.5) at a distance a little beyond that of Jupiter, and certainly not at the distance of the Earth's formation. The model that explains the lunar bombardment so well (Figure 5.2) predicts that 84% of the comets' mass came from Jupiter's zone, 14% from Saturn's, and only 2% from Uranus and Neptune. Hence it is consistent with an eight-fold enrichment of the D/H ratio present in our oceans. In contrast, only 8% of the Oort cloud comets originated in Jupiter's zone, 16% in Saturn's, 24% in Uranus' and 52% in Neptune's. Hence 'new' comets (which come from the Oort cloud) must often have a larger D/H enrichment ratio. It must be consistently even larger in short-period comets, because the Kuiper belt is a much colder place of origin than the giant planets' zones.

Thus, we now believe there was a bombardment of comets (that is, icy planetesimals 5 to 500 km in diameter) coming mostly from the zones of the giant planets. Their orbits were perturbed by the growth of those planets' masses and, for a billion years, they brought the water of the oceans, the gases of the atmosphere and the compounds of carbon needed for the emergence of life.

The atmosphere and the oceans of the Earth

The chronology given in Table 4.3 shows that the bombardment rate of the protoplanets fell virtually to zero in 200 million years; the last important collision with the Earth was perhaps that which formed the Moon. The number of the small planetesimals in the Earth's zone subsided in the same way; however, many small collisions with irregular objects between 1 and 20 km in size remained inevitable for a while.

Meanwhile, the cometary bombardment was continuing. It took 500 million years to come within 99% of completion. Its rate then fell to less than 1% of its initial rate, as clearly established by the lunar data. But on the Moon, gravity was too weak: each cometary impact dissipated into space all the water vapor and all the volatile matter; only the cometary silicates accumulated and added 2% to the Moon's mass. Through time, huge seas of lava eventually covered nearly 80% of the surface area of the Moon.

The size distribution of the lunar craters also confirms the size distribution of the impacting objects. The very large objects (50 to 500 km) were much rarer than the smaller objects, but they brought most of the mass.

The same bombardment struck Earth. The Earth's gravity enhanced the effect somewhat, but it had another important result: it retained most of the water and a good fraction of the volatile matter. Overall, however, the cometary contribution increased the mass of the Earth by less than 1%.

An important contribution of the cometary bombardment of the Earth came not from large fragments but from *ultramicroscopic dust*. As a matter of fact, those comets that come to the inner Solar System are sufficiently heated by the Sun to develop a dust tail (these spectacular dust tails are the reason for our historical interest in comets). An appreciable fraction of that dust remains stabilized in a flat disk that surrounds the Sun in the plane of the ecliptic. On a moonless night, it is still faintly visible when the Sun has just set, in places where the sky is very black. It is called the zodiacal light, because it is slanted in the direction of the Zodiac (i.e. the constellations along the ecliptic, which is the plane of the Earth's orbit around the Sun).

Needless to say, at the time when the cometary bombardment brought 1000 times more comets than now, that zodiacal light was 1000 times more intense. Since it extends to distances beyond the Earth, the Earth's orbit remains within it and the Earth constantly collects ultramicroscopic grains of this dust, which fall through the upper atmosphere. The U-2 stratospheric aircraft of NASA collected many of these 'interplanetary grains' that obviously come from comets. Analyzed by the American astronomer Don Brownlee and coworkers, these grains proved to be usually aggregates of submicroscopic particules, containing about 5% carbon and showing the existence of prebiotic organic molecules, unaffected by the gentle resistance of the upper atmosphere.

We may conclude that, during the first billion years of the cometary bombardment, some 10^{18} tons of prebiotic molecules were sprinkled over the whole Earth in the form of particles smaller than 1 μm; most of them sank into the oceans or were stopped by sea froth. This represents, on average, 200 g/cm^2 per million years. Of course, this *is* an average; the rate was much more intense during the first 500 million years, and subsided gently afterwards.

Let us now paint a picture of the Earth 4 billion years ago, i.e. 500 million years after its formation. A cloudy atmosphere scores of times thicker than the present atmosphere protects the ground and prevents it from cooling rapidly. This is important because the Sun, which first contracted while decreasing in total brightness, has just gone through a minimum during which it was 30% less bright than at the present time. From this moment onward, its brightness is going to grow very slowly, by about 7% per billion years.

On the Earth, most of the cometary water has turned into steam. This water vapor condenses to form the oceans as soon as the temperature of the ground falls below the boiling point of water (230 °C at 30 atmospheres). In the residual atmosphere, the solar ultraviolet and the high temperature induce photodissociations and polymerization of molecules. Formaldehyde (CH_2O) polymerizes (the process may even have started earlier in cometary dust), and a fog of solid particles of polyformaldehyde forms in the stratosphere. Ammonia (NH_3) is photodissociated into nitrogen (N_2) and hydrogen (H_2); the hydrocarbons oxidize and lose their hydrogen. This hydrogen escapes from the top of the atmosphere (the exosphere) and is quickly lost into space.

Less than 4 billion years ago, the ground might have already cooled down to 100 or 120 °C. The atmosphere then contains 80% carbon dioxide (CO_2), and perhaps still 10% methane (CH_4), 5% carbon monoxide (CO) and 5% nitrogen (N_2). This atmosphere has not been able to reach thermochemical equilibrium, in part because it is still fed by a diminishing cometary bombardment.

The atmosphere continues to be turbulent in the extreme. Incessant torrential rains of a water acidified with CO_2 attack the silicates on the ground. The ground temperature is soon low enough that the carbonates formed by this reaction become stable; they make solid sediments that bury the CO_2 in the ground and rapidly cause the atmospheric pressure to fall to maybe 10, then 5, atmospheres.

Three and a half billion years ago, the last large cometary impacts become rarer and rarer; but when they occur, they still produce enormous cataclysms. They induce local boiling of the oceans that causes hurricanes and gigantic waterspouts with fantastic ejections of gas and water into space. Under these chaotic and inhospitable conditions, a phenomenon occurs that is apparently insignificant. However, it is going to have astonishing consequences: bacteria multiply in the very hot waters of the first oceans.

Life has just emerged on the Earth.

The emergence of life on Earth

The first evidence that life had appeared on the Earth is probably present in ancient rocky sediments of Greenland that are 3.8 billion years old. This evidence is indirect. The rocks have been heated too much and have undergone metamorphism. For carbon, they show an enrichment of the isotope ^{12}C with respect to isotope ^{13}C; this enrichment could only have come from organic compounds found in living organisms.

It was in Australian sedimentary rocks that the American paleobiologist J. William Schopf found fossil remains of 11 different kinds of bacteria dating from 3.465 billion years ago (Figure 5.3). The variety present in 11 different groups of bacteria implies that these microbes had already had time to evolve sufficiently to be differentiated from each other. Their analogy with cyanobacteria (formerly called bluegreen algae) suggests that these microbes were already using solar light to dissociate CO_2 to make nutrients, and in so doing released half of the oxygen in the molecule into the atmosphere, just as cyanobacteria and plants do today.

The 3 billion years that followed provide us with numerous fossil relics called stromatolites. These are stratified rocks formed by carpets of bacterial colonies that were also using solar light. Stromatolite fossils can be easily recognized in the fossil rocks, from their similarity to those living stromatolites still surviving in the shallow waters of the Carribean and western Australia.

Seven hundred million years ago the first multicellular organisms appeared, leading to an acceleration of the diversification of species and their adaptation to new conditions. Trilobites and green algae appeared about 600 million years ago, soon followed by the first jellyfish. Three hundred million years ago the first aquatic plants and the

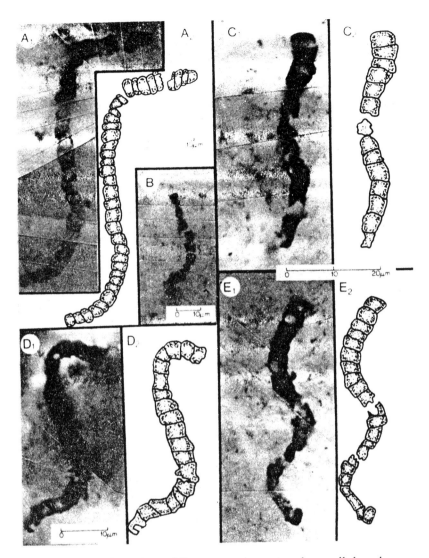

FIGURE 5.3   Fossil imprints of filamentous bacteria, whose cells have been preserved between layers of sedimentary rocks 3.465 billion years old in north western Australia. This is where 11 different kinds of microbes have already been discovered by the American paleobiologist William Schopf. These microorganisms are probably photosynthetic bacteria, somewhat similar to the cyanobacteria (still often called blue-green algae), implying nutrition by means of photolysis of CO_2, with the production of oxygen. The diversification and the evolution of these primitive species are surprising, since they occurred so soon after the cometary bombardment subsided.

first arthropods were ready to leave the sea, heralding respectively tree-like ferns and amphibians, then conifers and reptiles. Before discussing this evolution in detail, let us first try to define the nature of life.

Defining life

Too abstract a definition of life can easily lead to quarrels over semantics. Let us limit our task to an operational approach, and seek to define the fundamental differences that separated the first bacteria from the mineral environment that surrounded them. These differences are still seen in all forms of plant and animal life at the present time.

Above all else, the living organism is a system which allows both reproduction and evolution.

(1) *Reproduction* is a copying process that allows one bacterium to make another one that is identical to itself. In plants and in animals, this process, although more elaborate (we shall discuss sexual reproduction later), is not basically different. The seed of an apple becomes an apple tree that grows apples. The duck egg becomes a duck, cats beget kittens that become cats, a man and woman beget a baby that becomes a man (or a woman). This copying process is above all based on the transmission of the proper information, followed by the mechanisms needed to translate this information into chemical processes.

However, reproduction is not sufficient to define life. Indeed, when a crystal is immersed in a saturated solution of salt, it grows and reproduces the crystal structure of the neighboring atoms. A duplicator can copy as many copies as desired of a typed page. A living organism has within it a second process which differs from a simple reproduction; it is the process that leads to evolution.

(2) The *evolution* of forms of life is reached by an extraordinarily slow modification of the information transmitted in making a copy. It is evolution that provides a way out from the well-known dilemma 'Which came first, the chicken or the egg?' The existence of evolution allows an escape from the paradox. The chicken was formed by the slow evolution of an animal

being transformed through the course of ages, so that gradually one can trace it back to the primordial cells that emerged more than 3.5 billion years ago.

We now know that the teeming multiple forms of plant and animal life sprang from a single unique progenitor, by progressive evolution and complexity of the information transmitted to the eggs of the succeeding generation by the genes of the preceding generation. We are thus the more or less distant cousins, not only of chimpanzees and orangutans, but also of dogs and cats, of larks and ostriches, of oaks and roses, of ferns and mushrooms, of lichens and bacteria.

This is a statement which is extraordinary enough to require detailed justification; but before trying to justify it, we still have to complete our definition of life.

(3) What is required for the two processes of reproduction and evolution to work in living organisms? There must be a structure able to stabilize the series of necessary processes. These processes must also be supplied with the energy essential for their operations, and with the substances required to make the duplication. The structure must 'encode' a copy of the instructions, then 'read' them to transform them into another organism.

The basic structure is the cell, and primitive forms of life consist of only one cell. In contrast, plants and animals are multicellular; that is, formed of an often very large number of cells that are different and specialized in their various tasks.

Separation of the cell from the outside environment is secured by a thin semi-permeable membrane; it selects the molecules that may enter and leave, while maintaining a steady state that remains far from thermochemical equilibrium. Stability, although far from equilibrium, is provided by a complex system of mostly negative feedback processes which work in the same basic way as a thermostat for a central heating system. These processes give the cell a degree of autonomy from its environment. The substances required for duplication are brought from outside (nutrients) and the energy needed is produced by metabolic processes. These processes operate like a thermal engine, transforming the nutrients, often by oxidation, partly into work and partly into heat, while expelling waste outside.

Table 5.1. *The percentage abundances of the elements of life,*
compared with cosmic data

Element	Bacteria	Mammals	Interstellar frost	Volatile fraction of comets
Hydrogen	63.1	61.0	55	56
Oxygen	29.0	26.0	30	31
Carbon	6.4	10.5	13	10
Nitrogen	1.4	2.4	1	2.7
Sulfur	0.06	0.13	0.8	0.3
Phosphorus	0.12	0.13	—	(0.08)
Calcium	—	0.23		

Note:
This table also emphasizes that calcium was not present in significant
amounts in primitive bacteria; its use in shells and skeletons is a recent
discovery of the evolved animals. Phosphorus is not observed in comets; the
figure in parentheses represents its cosmic abundance, which is reasonably
expected to be present in comets. From these data, phosphorus was
concentrated by a factor of 1.5 in bacteria; presumably phosphates were
used as a source of energy before the cyanobacteria discovered the use of
photosynthesis.

We have not considered the question of the chemical composition
of living organisms. The six most abundant elements in all living
organisms are hydrogen, oxygen, carbon, nitrogen, sulfur, and phosphorus. Some 15 years ago, I compared the relative abundances of
these elements in living beings with that in the frost covering the
interstellar grains, and in the cometary snows. Table 5.1 shows the
results of this comparison; the table makes it clear that the composition of living matter resembles less that of the Earth (silicates of the
rocks) than that of interstellar frost or the volatile fraction of
comets!

The genetic code

So far, we have spoken of reproduction and evolution in general terms, as well of the cell as the unit of life; but we have not yet described the chemical code first used to write down, and then to copy, the genetic instructions for building a cell. During the last 50 years, this chemical code has been completely elucidated.

The instructions are recorded in the DNA (deoxyribonucleic acid) which embodies the lasting memory of the species, and is faithfully copied and transmitted to each generation. DNA is a molecular strand of comparatively gigantic length (in a human cell, about a meter), located in the cell nucleus of higher cells and the cytoplasm of bacteria; this strand is protected chemically, during nearly all its life, by a 'negative' copy of itself. These two complementary strands are wound together in the shape of a double helix, like the two threads of a twisted string (see Figure 5.4). The twisted coil is then wound upon itself in order to remain compact.

A single specimen of this duplicate copy lies inside each of the cells of a multicellular animal, just as in the single cell of the bacteria. In human beings, the DNA is large enough to contain the amount of information equivalent to 100 million words (a hundred thick books). Fragmented into 24 pairs of chromosomes, the strand holds the recipes to make thousands of proteins of different types. All these proteins have a specific function; not only do they supply construction materials for the body (skin, muscles, hair, cartilages, bones, etc.), but also the workers, the architects and the master plan (enzymes, catalysts that select some chemical reactions or prevent others).

DNA fragments are 'transcribed' (that is, copied) by another nearly identical molecule, messenger RNA (ribonucleic acid). The RNA in higher cells then comes out of the cell nucleus to be 'translated' into a specific protein, through a biochemical mechanism sketched in Figure 5.5.

Let us limit our description to the essentials. DNA contains a coded message through the linear use of four 'letters' in a variable order. These four 'letters' are the bases adenine, thymine, guanine and cytosine, or symbolically, their four initials A, T, G and C.

Transcription is accomplished by 'pairing' the bases in accordance with their geometry (see Figure 5.6). The form of base A fits base T

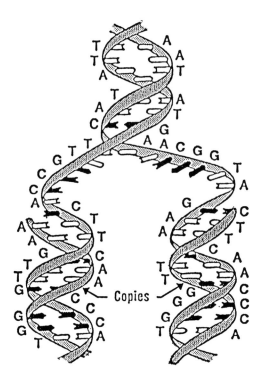

FIGURE 5.4 Schematic diagram of the DNA double helix, and of its copying process for duplication. The strands in the helix are formed by a chain of alternative molecules of sugar (pentoses) and phosphate, not shown here. The horizontal steps are formed by pairs of nucleic acid 'bases' (A·T, T·A, C·G, or G·C), whose order along the strand encodes the genetic information. To make a copy, the pairs of bases separate temporarily.

(and not C or G). When they are sufficiently close, bases A and T attach to one another by two weak bonds called hydrogen bonds. Similarly, base G accepts base C and refuses A and T; when very close, bases G and C attach to one another by three hydrogen bonds. It follows that the pair G·C forms a stronger bond than the pair A·T.

Instead of the thymine present in DNA, RNA uses uracil (U) which acts in the same manner to form the pair A·U. The only other

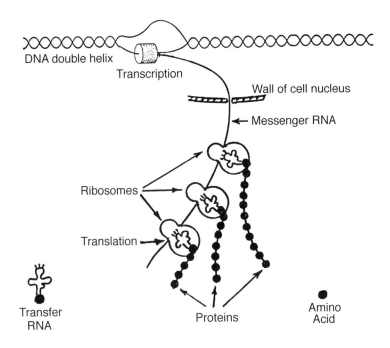

DNA double helix

Transcription

Wall of cell nucleus

← Messenger RNA

Ribosomes

Translation

Transfer
RNA

Proteins

Amino
Acid

FIGURE 5.5 Sketch of the mechanism of assembly of amino acids into proteins in nucleated cells. DNA is a very long twisted double strand, which contains thousands of recipes for the assembly of many types of proteins from only 20 types of amino acid. Each recipe is written in 'words' of three letters (three nucleic acid bases chosen from among the four A, T, G, C possibilities). Each recipe is cut off from the following one by a 'punctuation mark'. It is transcribed by a piece of messenger RNA, which comes out of the cell nucleus and meets ribosomes, the tiny machines that assemble amino acids into proteins. The ribosome is itself part RNA and part protein. It slides along the messenger RNA and 'translates' it, while using numerous different 'transfer RNAs'. The transfer RNA is a piece of RNA folded as a clover leaf, which carries at one end three unpaired bases (one 'word') and, at the other end, the amino acids that define this word. The ribosome acts as a lock for the keys of the transfer RNAs which come one by one and each add a 'word' – the correct amino acid for the protein which is forming.

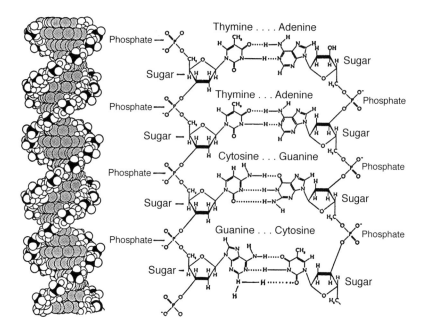

FIGURE 5.6   The chemical structure of DNA (deoxyribonucleic acid). On the left is the shape of the double helix, with the different atoms identified: in white, H, O and C in the phosphates and sugars; in gray, C and N in the nucleic acid bases (horizontal steps in the ladder); P (phosphorus) is in black. On the right is the structural formula. Between the two identical vertical chains which alternate sugars (pentoses) and phosphates, the *bases* form horizontal steps; they are linked by pairs, thymine with adenine (A·T) and cytosine with guanine (C·G); the pairs of bases used have exactly the same length. The dotted lines represent hydrogen bonds; they result from the quantum property of *electron sharing*. In RNA (ribonucleic acid), a single hydrogen atom is substituted for the radical methyl group (CH_3) of thymine (the base no longer being called thymine, but uracil). The symbol C for carbon has been omitted from all the cyclic molecules to simplify the drawing. The twisting into the shape of a helix is due to the conformation of the sugars whose pentagonal structure is not in the plane of the figure; its spatial orientation is suggested by the bolder printing of the bond on the side nearest the observer.

difference between DNA and RNA is that the sugar molecule (see Figure 5.6) is not the same.

The thread that joins the bases and determines their order can be compared to the thread that holds a necklace together. It consists of a regularly repeated combination of identical sugar and phosphate molecules. The three-dimensional geometry of these molecules is such that these two 'pearl necklaces' tend to wind with a twist that makes the double helix. They are separated by the pairs of bases which attach them together, like the steps of a spiral staircase.

Lastly, the translation into the language of the proteins is made through the use of three 'letters' per 'word'. The three letters are taken from the four-letter 'code' (A, U, G, C) and a three-letter word defines one amino acid.

There are 64 possible ways to arrange four letters in groups of 3, but a number of them are synonyms, so that they are limited to 20 different amino acids plus three punctuation marks, needed to cut the message at the right place. The 'proteins' are assembled as strands of some scores or hundreds of 'words', which select the amino acids in a precise order.

The structure and three-dimensional geometry of each protein is therefore written in the initial message carried by the DNA, i.e. the pairing of the bases A with T and C with G, which offers only two binary choices: AT or TA, CG or GC. That choices so limited could lead to structures so complex and so varied as those of all the living beings is not really surprising, since it is how computer programs operate. What is surprising is that it has developed spontaneously and we must investigate how it evolved.

First, the great complexity comes exclusively from the order in which the molecules are assembled. Only five nucleic acid bases, plus two types of sugar molecule and one phosphate are needed to build DNA and RNA, and only 20 different amino acids to build an immense number of different proteins. Let us add water and a few lipids for the cell membrane; all of this requires not much more than 30-odd different molecules. Most of these molecules are built up using H, C, N, and O, by far the most abundant atoms in the Universe (with the exception of helium, which is chemically inert). In addition, only two of our 30 or so molecules contain S, the sulfur atom, which is also among the 10 most abundant elements. Only one, phosphate, contains P, phosphorus. As mentioned before, phosphate was probably the first source of energy of the primitive cell.

This complicated hierarchy of chemical structures that issues from only a few very abundant atoms, explains the whole variety of living forms and their complex evolution.

Fossil traces of evolution

The study of the fossils found in geological layers has allowed a partial reconstruction of the evolution of the animal and plant species that took place during the last 600 million years. The different geological epochs have been dated with accuracy, thanks to radioactive isotopes. At the beginning of this century, geologists still believed that there was no fossil more than 600 million years old, but this was because they had not yet persisted in examining more ancient geological layers with a microscope. We now know that bacteria preceded multicellular animals and plants by about *3 billion years*!

Of course, it becomes more and more difficult to identify fossil fragments in more and more ancient rocks, because they have undergone profound geological transformations. Moreover, *sedimentary* rocks are needed in order to imprison and fossilize living remnants of animals or plants in the sediments, and sediments older than 3.5 billion years are seldom found. Now we know why: the cometary bombardment was too intense during the first billion years after the formation of the Earth. As already mentioned, the most ancient known fossil bacteria are from Australia and are 3.465 billion years old.

To make it easier, we shall speak hereafter of ages as dating back from the present. We have used for age zero of the Solar System, the time when the dust separated from the gas in the accretion disk surrounding the early Sun. The radioactive ages of many chondrites are in good agreement with a date for this age zero of 4.56 billion years ago. The Earth had probably reached 98% of its final mass 4.50 billion years ago, and the cometary bombardment remained very intense up to 3.8 billion years ago, when the Earth had reached 99.9% of its final mass.

In this epoch, the large cometary impacts were separated by longer and longer periods, so that their effects on the climate were no longer cumulative. Since most impacts influenced only a small fraction of the total surface of the Earth, it is likely that colonies of microbes were able to survive in at least one location of the globe, and propagated from there about 3.8 billion years ago. This seems to be confirmed by the isotopic enrichment of the carbon isotope ^{12}C found in some sedimentary rocks of Greenland.

However, understanding of how the multiple forms of life have emerged and developed on the Earth was helped not so much by the geological study of fossils as by the recent techniques of molecular biology.

Evolution of the genetic message

I have already emphasized the importance of evolution as one of the two processes used to define the phenomenon of life; but its mechanism has not yet been clarified.

The process is extraordinarily simple, because it comprises two principles:

(1) Errors in copying the genetic message, which give rise to *'mutations'*.

(2) Survival of the fittest; that is, of the best adapted to the environment.

The first principle rests on the impossibility of avoiding errors in transcribing the genetic message; all in all, it is just like typing mistakes or misprints. A typical copying error of the genetic message comes from misreading a 'punctuation mark', e.g. a 'stop' signal. The message may be copied a second time before meeting the stop sign again. In this case, two identical sequences will be passed on in different cells to future generations. In the future, they will become more and more different because they will accumulate different copying errors in the two segments.

In the duplication of DNA (in its 'replication', to use the biological term), the rate of copying errors is remarkably constant and is about one error per billion pairs of bases copied. For a typist, this would correspond to a single typing error per half million pages! Such accuracy is due to the existence of a correction process that removes the errors and 'repairs' the message, in the way that a careful proofreader can find and correct almost all the misprints.

However, the length of the molecular strand to be copied is immense. Human DNA amounts to more than 10 billion pairs of bases, so that there must be about a dozen mistakes per copy, each time a human cell is duplicated. The resulting differences give rise by and large to insignificant mutations. This explains why our children sometimes have features they have not inherited from either of their parents.

One consequence of those copying errors that skip punctuation is that the message transmitted to further generations may always be a bit longer but never shorter. This has constantly occurred throughout the course of time. In Figure 5.7, the length of the DNA strand of an individual species is plotted against its epoch of emergence; the DNA length is measured by the amount of information it can carry – to describe it more intuitively, information has been translated into the number of 'words' present in the message. However, because the number of times the 'stop' sign has been skipped is numerous among plants and animals, there are long portions of the message that have remained incoherent, and are not recognized by the RNA during the transcription process (see Figure 5.5). These long fragments of DNA that are not used to make the plant or the animal are called 'introns' or sometimes 'junk DNA' by molecular biologists, and those that are being used are the 'exons'.

The introns do not remain useless. During evolution, a new copying error by the molecular mechanism that discarded them may make them reappear, sometimes to create a monster or at least a deformity, but other times to create a feature fitter for survival in an environment that has changed.

Figure 5.7 shows not only the race toward complexity of evolved organisms, but also carries an important message with regard to the origin of life. It suggests that the earliest forms of life began evolving with an extremely short genetic message, which implies that the first living beings were extraordinarily simple, probably much simpler still than a virus. We know that today a virus behaves parasitically: it can only survive and multiply within living cells, but it possesses a simpler system of replication than the most primitive cells. Its genetic recipe is written either in DNA or in RNA, but it never combines both systems. The simultaneous existence of DNA (to carry the genetic message) and of RNA and ribosomes (to translate the DNA 'letters' into the amino

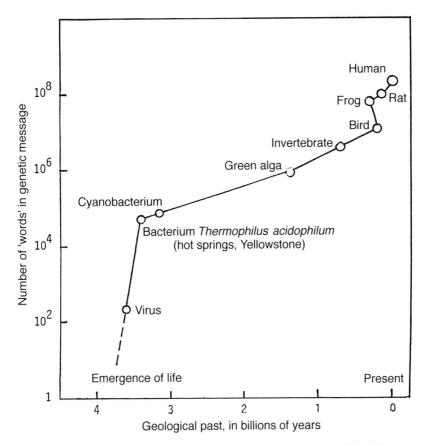

FIGURE 5.7   The maximum amount of information that could hold the genetic message of different living organisms, as a function of their emergence date on Earth, in billions of years. The dizzying growth in complexity is displayed for the evolved animals; it also shows that, at the emergence of life on Earth, the genetic message had to be extraordinarily short (from one to ten 'words'); I have used 'words' here to help to visualize the quantity of information. A five-letter word of the usual language holds about 100 bits of information (1 bit = 1 choice between 2 possibilities). One could argue whether the viruses should be included here because they are only 'parasites' of the bacteria, but they must have the approximate features of the unknown 'protobiont' that was the ancestor of the bacteria. The amount of information actually used by plants and animals is somewhat less because of the existence of 'introns' (see the text).

acid 'words') in the most elementary bacterial cells, already shows a big step up on the climb toward complexity.

Finally in Figure 5.7, the slope of the curve from viruses to bacteria is steeper than that from bacteria to animals and plants, because the RNA editing mechanism that cuts out introns is not yet in place, hence the error rate is much larger. Evolution for the simplest animals works much faster. If you need to be convinced, think of the 'flu virus, which changes from year to year, eluding effective vaccines – to our great detriment.

Survival of the fittest

We still have to consider the second principle basic to evolution: that of the survival of the fittest, i.e. of the individual best adapted to its environment. Such an individual has some features that give it an advantage in its competition with other individuals of the same species. The nature of this advantage matters little, whether it be in the search for food, or in protection against a predator, or against a change of climate, or for any other reason. The insignificance of this advantage also matters little, provided that more of the individual's offspring survive. What matters is that this advantage is going to increase the relative number of descendants in *geometric proportion* over many succeeding generations.

A numerical example may be useful. Let us imagine that the available food limits the population of an ecological niche to 10 000 individuals. If a mutation gives an advantage of only 1% to a single individual, in 1000 generations, its descendants will have eliminated *all* the other individuals. And 1000 generations is a blink of an eye in terms of geological time.

It was the English naturalist Charles Darwin who proposed for the first time the correct mechanism that explains how a species can change: he called it 'natural selection'. His book *On the Origin of Species by Means of Natural Selection* was published in 1859. It was not readily accepted because it ran counter to common prejudice. Darwin had observed with diligence all the details of the variations from individual to individual within a species, which caused the species to be progressively shifted from its initial form. Of course he had no idea of the origin of these changes; we now understand they come from mutations caused by copying errors of the genetic message. The genetic studies of the Austrian botanist monk Gregor Mendel, followed by those of the American Thomas Hunt Morgan, clarified one important part of the problem, and what was called the 'Neodarwinian synthesis' asserted itself at the beginning of the twentieth century.

However, it was still fraught with difficulties in explaining gaps in the continuity between animal fossils, presumably due to the lack of sufficient fossil traces. Thus it is not the study of geological fossils but rather the molecular biochemistry of still living species that allowed sudden and substantial progress in this field.

At the beginning of the twentieth century, the neodarwinists introduced the concept of the 'gene' as a discontinuous unit for the transfer of heredity. Morgan established that genes were localized on the *chromosomes* present in the cell nucleus. From 1944 on, it was demonstrated that it was DNA and not proteins that carried the genetic information. Finally, the British biophysicist Francis Crick and the American James Watson discovered the molecular structure of DNA (the double helix) which opened the way to elucidating the mechanisms and processes of information transfer. The amino acids are assembled in a specified order by the DNA to make a protein. The 'gene' is thus the fragment of genetic information localized in the DNA and copied first by messenger RNA and then via molecules of transfer RNA to allow the assembly of a protein (Figure 5.5).

As soon as it was known how the order of the amino acids in a protein was decoded, it became possible to compare that order in proteins which have a similar function in different animals. The frequency of the sequences of amino acids that were identical or different in any one protein group could then be used to compare the degree of common ancestry among the species.

Evolution in molecular biology

The amount of information contained in the genetic message is prodigious, as already shown in Figure 5.7; even the introns (regions not used) are useful for our comparative studies. But we have not yet been able to read the whole of the message present in the DNA of higher animals, quite simply because of the immensity of the task. In different animals, however, it is possible to identify long sequences that match, in the same way as it would be possible to identify long identical sentences in two editions of a book, without having to read the whole book.

In order to reveal the complexity of the genetic message in the most primitive organisms, let us look at simple bacteria: they have no nucleus, only one chromosome, the genome (which is circular: closed on itself, with no beginning and no end). However, their DNA contains about 3 million pairs of bases (A·T, G·C) that can be transcribed into about 2000 different proteins, each containing on average a 'sentence' of 500 'words' (written in the vocabulary of the amino acids).

In order to embark on comparative studies, it is enough to choose a single protein and to compare the order of its sequence of amino acids in different animals. This was first started with an easily identifiable protein, hemoglobin, which colors the blood red and whose function is to carry oxygen through our circulatory system. The study of many hemoglobins from different animals has been extensive; it was possible to construct a genealogical tree for hemoglobin which shows, in partic- ular, that two protein chains (called alpha and beta), both present in hemoglobin, came from the duplication of a single ancestral chain found in the lamprey, a primitive fish without either jaw or limb. The hemoglobin of the lamprey consists of a single chain of molecular weight 17 000.

Let us consider another example: that of the genealogical tree of cytochrome c, an essential pigment of cellular respiration. This genealogical tree (or more exactly, this phylogenetic tree, to use the ter- minology of the molecular biologists) is displayed in Figure 5.8. It shows that the recipe for the pigment in question has been pro- gressively modified by mutations, so that the percentage of differences in the order of the amino acids of the pigment shows the degree of common ancestry between species. For instance, Figure 5.8 shows that the cytochrome c of the dog and of the rabbit differ by only 2%.

To work out the ancestry of two more distant species, one needs only to follow the shortest path between them. For example, to compare sunflower with wheat in Figure 5.8, one adds the numbers $6 + 4 + 8 = 18$, which shows that 18% of the amino acids present in the cytochrome c of these two plants are not arranged in the same order. The pigment of wheat differs by 8%, and that of the sunflower by 10% from that of their common ancestor.

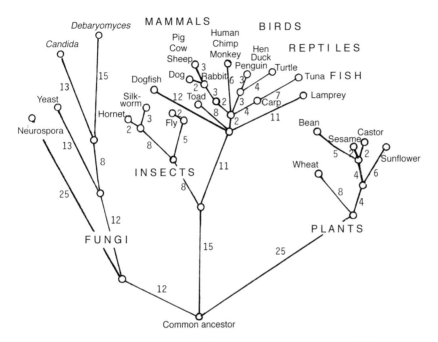

FIGURE 5.8   The phylogenetic tree of cytochrome c (a protein which is an essential pigment for cellular respiration). Each circle represents the cyctochrome c of a well-defined species. The numbers between two circles are the percentage differences in the order of the amino acids of cytochrome c in two different species. The lowest circles represent the common ancestors, often no longer extant, for the lines that diverge upward. The mutation rates due to transcription errors are more or less constant and are approximately known; hence the diagram allows for computing the approximate age of the common ancestors.

One can also try to reconstruct a phylogenetic tree that leads back to the bacteria. In each animal, or plant, or fungus, or bacterium, a very large number of different proteins is available. The only difficulty, but it is huge, centers on the immensity of the task of reading the 'words' of each 'sentence' of the code (if one delves into the amino acid order of the proteins) or even the 'letters' of each 'word' (if one deals with the DNA code). It takes immensely longer to read the whole string of DNA than one or several proteins, but one needs only to find nearly identical 'pages' in the undecoded 'books'.

The American biochemist Carl R. Woese has probably done most to elucidate the genealogy of bacteria, whose importance is obvious when we take a long look on the distant past. Woese chose the RNA present in ribosomes, because they can be used as a clock for the evolution of organisms that have only a rudimentary mechanism for translating DNA into proteins. The ribosome is the mechanism by which the DNA letters are translated. First transcribed by messenger RNA, these letters are read in groups of three and translated into 'words' (amino acids) by transfer RNA and the ribosomes, which align them in 'sentences' (the protein sequence of amino acids).

In contrast to the DNA whose double helical form is quite simple, the ribosomal RNA has a complex structure, hairpin-folded about 50 times upon itself. These 'hairpins' are stabilized by complementary bases that attach the two branches to one another (see Figure 5.9). The RNA part of the ribosome of the bacterium *Escherichia coli*, for example, consists of more than 1500 bases. The distribution of these bases in many bacteria allowed the building of a genealogical tree, comparable to the one for animals, plants and fungi shown in Figure 5.8. This gigantic task took many years. However, when it was combined with data from other proteins, it established a 'universal tree of life'. Figure 5.10 suggests only part of this (its true extent would be too detailed for our purposes).

Figure 5.10 also shows that there was a common universal microbial root for all forms of life. The trunk first split into three branches, called Archeobacteria, Urkaryotes and true bacteria. From the Urkaryotes, the first eukaryotic cells formed (cells with a separate nucleus); then, plants, animals and fungi appeared. The first eukaryotic cell emerged about a billion years ago, so that, for almost 3 billion years, there were only prokaryotic bacteria in the biosphere.

Figure 5.11 has only some 30 branches, although just over 2 million of them are known. Woese's critical examination showed that it is difficult to determine with certainty the position of the common ancestral bacterium; indeed some scientists still question whether there was one only. Life could have emerged a large number of times in different locations in the biosphere, but it may very often have been destroyed, first by the great cataclysms, later by competition and survival of the fittest. The essential is that one single root for extant organisms is not unlikely.

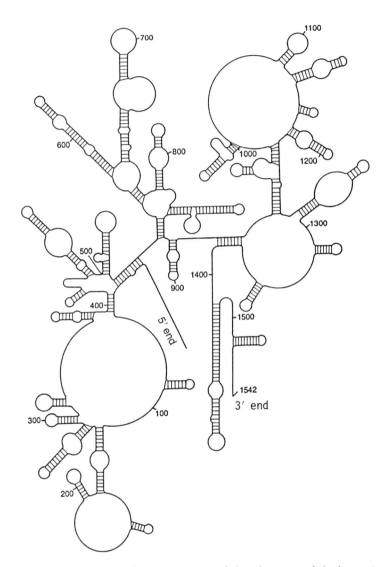

FIGURE 5.9   Structure of the RNA part of the ribosome of the bacterium *Escherichia coli*. The ribosome is a tiny structure present in many copies in each cell. Its function is to translate the DNA (via fragments of messenger RNA) into an ordered sequence of amino acids (a protein). In this remarkable 'translation machine', the changing structure of the nucleic acid bases provides a 'clock' for the evolution that enables one to work back to the most primitive forms of bacteria.

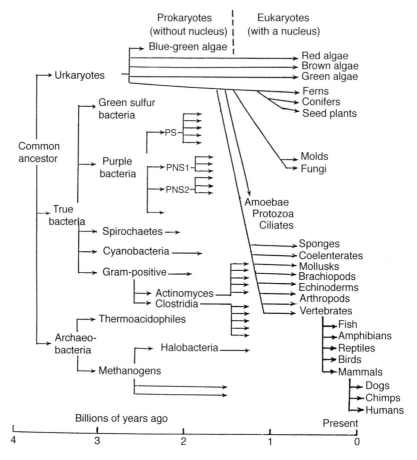

FIGURE 5.10 For close to 3 billion years, life on Earth was limited to microbes: bacteria and other monocellular organisms. Non-microscopic animals and plants emerged in the seas 700 million years ago, and came out from the sea less than 400 million years ago. The branches that diverge from the common root or ancestor are based on Woese's interpretation; he used analysis of the RNA sequences in ribosomes as a universal clock for evolution. The position of the common ancestor remains somewhat uncertain, but its existence can no longer be in doubt.

Our problem is now to understand how life was really able to emerge on Earth some 3.8 billion years ago, i.e. as soon as the intensity of the cometary bombardment began to fall off significantly. The puzzle remains hidden in the 'dawn of time' or, more exactly, in that lost interval of the first billion years that we now assign to the enormous plowing of the terrestrial crust by comets. We are reduced to speculation, but recently this has been supported so well by solid biochemical and astrophysical evidence that the question is in the process of being rapidly clarified.

The origin of life

In 1924, the Russian biochemist Alexander Ivanovich Oparin was the first to consider the problem of the origin of life in modern terms, followed in 1928 by the British biologist John B. S. Haldane. In 1947, the British physicist John D. Bernal suggested the importance of clay surfaces in the polymerization of prebiotic organic molecules. In 1951, the American chemist Harold C. Urey (Nobel laureate for chemistry in 1934 for his discovery of deuterium) discussed the importance of Oparin's idea that the primitive atmosphere of the Earth not only could not contain oxygen, but had to be reducing (with a hydrogen excess) to help the first chemical reactions leading to life. Using these ideas, one of Urey's young graduate students, Stanley Miller, undertook a laboratory experiment that would become famous and initiate a new start in the research on the origin of life. It was published by Stan Miller in 1953.

Miller's experiment consisted of exposing a mixture of methane (CH_4), ammonia (NH_3), and water vapor (H_2O) to an electrical discharge in the presence of an excess of hydrogen. The mixture was constantly recycled and, after a few days, the condensing water became more and more of a red-brown color, containing many new, more complex organic molecules. In particular, there were four of the amino acids normally present in proteins: glycine, alanine, aspartic acid and glutamic acid. This experiment showed for the first time that, with a hydrogen excess, amino acids could form spontaneously from the simplest molecules, provided of course there was no oxygen. In order to synthesize them on a sterile Earth, it was assumed that the electrical discharge was lightning during thunderstorms.

In the next 30 years, Miller's experiment was verified many times, with diverse initial molecules and by replacing the electrical discharge with ultraviolet light. Most of the possible paths that may lead to the other molecules required by life have also been explored. The present situation is summarized in Figure 5.11. Whatever the initial conditions, the synthesis of the amino acids implies, by and large, a route via hydrocyanic acid (HCN) and formaldehyde (H_2CO) (this is what the chemists call the 'Strecker synthesis'). Likewise, it is easy to make purines and pyrimidines from HCN and substituted molecules such as CH_3CN.

Most of the syntheses on the top part of Figure 5.11 seem to exist in interstellar space, as mentioned in Chapter 4. Water is the most abundant triatomic molecule in the Universe. However, no liquid water can exist in space, and liquid water seems to be required for the poly-condensations at the bottom of the diagram. The last syntheses seem to imply alternate wet and dry cycles, which assumes the alternate presence and absence of liquid water (as in a tidal pool). Since the top of the diagram works in interstellar space, and its products were brought to the primitive oceans by cometary dust, Figure 5.11 suggests a straightforward scenario for the appearance of life on Earth.

The metal ions mentioned are essential, although needed in very small quantities, for some proteins that are 'enzymes' (catalysts). Figure 5.11 summarizes the ways and means for bringing together the various chemicals required for life, but it does not provide any indication of the process for assembling them into a working mechanism, nor does it tell us whether some precursors were needed in order to assemble the first cells, and make them work. To clarify this question, we must fall back on general principles. Life essentially relies on what the Belgian physical chemist Ilya Prigogine (Nobel laureate for chemistry in 1977) has called a 'dissipative structure'.

Dissipative structures

A dissipative structure is a chemical system maintained and organized outside thermochemical equilibrium. It seems to defy the second law of thermodynamics, which asserts that *entropy* (which is a measure of the unusable energy) can *never* diminish in a *closed* system. But a dis-

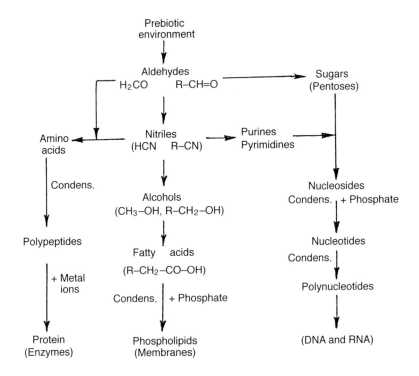

FIGURE 5.11   The chemical reaction pathways that led to the three groups of molecules needed for the emergence of life. Amino acids are required to make proteins; many proteins are enzymes, i.e. catalysts that select and direct chemical reactions. Phospholipids form semi-permeable membranes that build the cell walls. Finally, nucleic acids (DNA and RNA) are required for the hereditary transmission of genetic instructions. Liquid water is apparently needed for the polycondensations ('Condens.' in the figure). It is very probable that life emerged in water.

sipative structure is an *open* system that is constantly crossed by a flux of matter and energy, which permits entropy to diminish and the system to become organized.

Prigogine showed that the key to the processes present in dissipative structures lies in the non-linearity of the systems held out of equilibrium. Such systems can undergo a series of forks that branch off toward a more complex internal organization, and a consequent decrease in their entropy. Let me try to express this statement in more concrete terms.

The speeds of the chemical reactions are approximated by linear laws for most cases. But Prigogine and coworkers have systematically explored the behavior of non-linear systems. Such behavior appears when there is a feedback loop. We are all familiar with the feedback loop controlling our central heating systems; a change in the temperature induces a change in the amount of heat produced. There is a feedback loop here because a change in the effect is fed back to modify its cause. Any feedback loop may give a false sense of finality, as if an unthinking mechanism 'knew' how to influence the future. In chemical systems, when the feedback loop is strongly non-linear, as in autocatalysis, activations or inhibitions of reactions, this leads to transitions into a hierarchy of more and more complex states.

The German chemist Manfred Eigen, Nobel laureate in chemistry for 1967, tried to express the same idea in a more concrete language for the origin of life. He took the Darwinian idea of the survival of the fittest back to the level of prebiotic chemical reactions. In this case, it means the survival of the molecules that are the most apt to reproduce on a template, probably mineral-based at first. Eventually, such a template, by switching progressively to organic molecules, may lead to a ladder of greater and greater complexity.

The present discussions among scientists concern the identification of those molecules that are the most apt to reproduce. The French biochemist André Brack thinks that short strings of amino acids were able to form peptides first (a peptide is a very short protein, made of a few amino acids only), before the emergence of RNA. These molecules have to be organized through competition so that the most apt to reproduce will be selected.

The discovery of the fact that RNA exhibits autocatalysis has focussed attention on the 'RNA world' that may have come before the DNA world. Autocatalysis means that RNA can modify itself spontaneously. The Canadian Sydney Altman and the American Thomas Cech discovered that ribozymes (or enzymes made out of RNA) are autocatalytic; this discovery brought them the Nobel prize for chemistry in 1989.

According to the British biochemist A. Graham Cairns-Smith, there is still a missing link because evolution has not always depended as now on the close association between nucleic acids and proteins. He believes that the first organisms were a class of colloidal mineral crystals that formed steadily in open systems, while already following the rules of natural selection and evolution. Some of these primary organisms, already evolved, would have begun to make organic molecules by photosynthesis, using solar light. Gene control would have passed only gradually to amino acids. Life would have started in clay-like minerals, and the phosphates would have helped in the transition toward organic molecules.

The Belgian Christian de Duve, Nobel laureate in medicine in 1974 for his discovery of lysosomes (digestive organelles in the cell), also thinks that the autocatalysis of RNA is too limited and too slow to explain competition in the sense of Eigen. The major problem for a primitive RNA is the requirement to attach a phosphate to the purine or the pyrimidine, before polymerizing the nucleotide in order to make a strand of RNA.

Like Brack, de Duve believes that the set of theoretical and observational data implies that proteins were born as small polypeptides, with a length of 10 or at most 12 amino acids, and that they were selected one by one for their function, whether structural or catalytic. Central to Christian de Duve's original idea is the role of the thioester link, because of its high energy level, and of the existence of water acidic enough to dissolve the calcium phosphate present in the rocks. Thus the 'thioester world' could be the missing link still hidden in the 'dawn of time'; we shall probably soon reconstruct this missing link, from all the fossil traces still hidden in the DNA, RNA and proteins of living organisms.

I have given an incomplete glimpse of studies under way, but my goal was to convey only the view of the majority of the involved scientists: the remaining difficulties in explaining the origin of life from inert matter arise only from our ignorance, and not from apparently blind alleys which would all turn out to be dead-ends.

Chance or necessity?

Was the emergence of life unavoidable? Is it the result of a process that would have had to occur sooner or later? Or else is it the outcome of coincidences so improbable that time spans much longer than the age of the Universe would be insufficient to explain it by a random process?

Of course, when one has plenty of time, even the improbable becomes possible. When one plays dice for a very long time, one always ends up by throwing a double six three times in a row. In cards, one ends up dealing out four aces. Coincidences that are still more improbable will occur one day, if given much more time to happen.

The theory of the evolution of species initially gave rise to much disbelief. In the nineteenth century, this was in part because geological periods were believed to be much shorter. Today, it is probably because the disbelievers cannot imagine the immensity of the time lapse corresponding to four billion years. However, the recent phylogenetic studies (Figures 5.8 and 5.10) have convinced all serious scientists that evolution is no longer in doubt. The most important unsolved problem is the time span needed to produce the first bacteria. Between the decline in the number of large cometary impacts and the existence of the first fossil bacteria there cannot be much more than 200 million years. In our ignorance of the mechanisms needed to assemble the first self-reproducing cell, can we assume that a sufficient time has elapsed? Most biologists answer in the affirmative. But in 1970 the French biochemist Jacques Monod (Nobel laureate for medicine in 1965) wrote a widely read book entitled *Chance and Necessity* wherein he expressed the opinion that the origin of life was the result of a very unlikely chance that would not repeat itself elsewhere. However, Monod was unaware of all the more recent evidence that suggests otherwise.

Other scientists have addressed the residual problem of the time span needed to put together the first bacteria. In particular, the well-known British astrophysicist Fred Hoyle and his Sri-Lankan collaborator N. Chandra Wickramashinghe have endeavored to demonstrate that it is utterly impossible to build a bacterium in so little time. This is a useful effort, because it compels us to reexamine the situation. They started from the three-dimensional form of an enzyme, which gives it its specificity (for instance, its ability to fit the shape of another molecule in the way a key fits a lock). They computed that there is a 1 in 10^{15} chance of assembling amino acids by chance to build the required geometric solid. Then there is at most 1 chance in 10^5 of positioning the active site of the catalyst at the best location. Thus, there is only 1 chance in $10^{15+5} = 10^{20}$ of obtaining the required enzyme that is able to function. Trying to assemble it by chance at the rate of 1000 combinations per second would require 3 billion years!

This is not the problem, because it could be tried in a billion different locations in the early seas. Only three years would then be required for the enzyme to form by chance in at least one location. The problem lies in the fact that, in order to make a basic bacterium, about 2000 enzymes are required, each having a specific shape and a different catalytic action. In order to make them all by chance, at least $10^{20 \times 2000} = 10^{40\,000}$ trials are needed. This is a number so large that more than 10 pages would be required to write the 40 000 zeros which follow the digit 1. Fred Hoyle says quite correctly that the chances of such an event are comparable to the chances 'that the whirlwind assembles a Boeing 747 from scraps in a junkyard'.

The huge improbability of assembling a bacterium *by chance* within the 100 or 200 million years available on the primitive Earth is used by Hoyle and Wickramashinghe as the starting point for their hypothesis that bacteria preexisted in comets.

Hoyle and Wickramashinghe used my Table 5.1 as a further argument. Sir Fred writes in *The Intelligent Universe* (1983, p. 76, Holt, Rinehart and Winston, NY) that:

> Professor Delsemme has concluded from his results that cometary material must be the feedstuff of life. It is a conclusion that might seem remarkable enough, but one which I think is too cautious. Cometary material is life, I would say, not only its precursor.

Let me add that Hoyle and Wickramashinghe have tried to bring in many arguments in support of their hypothesis, but none is sufficiently decisive, so that the scientific community has not been convinced. Moreover, we must return to their probability assessment to show where it is misleading.

As a matter of fact, their computation is based on pure chance. They attach equal probabilities to all possible cases, which comes down to constructing a bacterium all at once from nothing. Despite its small size, it is true that a bacterium would be nearly as complicated to build as a Boeing 747, but there is no reason to assemble either a Boeing or a bacterium in a single operation. Of course, the difficulty lies in the nature of the evolutionary process, when all evolving proteins must help one another and act in concert for the sake of the whole organism.

Recent researches have begun to elucidate this question; they seem to imply that the emergence of life comes from the collective properties of the polymers that show catalytic features. This explains the recent interest in the fact that RNA is autocatalytic. Eigen has recently helped to clarify the problem. He has studied the way in which what he calls chemical 'hypercycles' can evolve. The main point of a hypercycle is that it involves several chemical reactions coupled by several feedback loops acting upon one another. I mentioned earlier that the simplest feedback loop (e.g. in a central heating system) gives a false sense of finality. When many feedback loops are coupled, this feeling is even more marked, because the system seems to 'know' how to influence the future. Chemical hypercycles have now been studied in the laboratory, and their spontaneous evolution in the course of time shows natural selection in action, as if it 'knew' the way to build more stable and more complex hypercycles.

Let me outline the simplest possible hypercycle. Three reaction products A, B and C act indirectly upon each other. A catalyses the reaction leading to B, B does the same for C, and C for A, closing the feedback loop. If some of the reaction products A', A'', A''' etc. are less fit to survive, they trigger the disappearance of their whole cycle. The same happens for B', B'', B''' etc. and for C', C'', C''' etc. If A survives, this is because it has selected the right choices for B and C, and vice versa. The hypercycle is therefore a machine for selecting and encoding the proper information for survival at least expense, and beginning with zero information. In a word, it is the Darwinian mechanism of evolution of the species and of survival of the fittest that has just been moved back to the level of a purely chemical process.

Laboratory experiments have now shown that the short period of time needed to begin early life on Earth is not a problem. With a simplified enzyme including only a few amino acid residues and maybe some fragments of RNA, a hypercycle could have produced the first 'protobionts' (Figure 5.7) in much less than 100 million years. Thus the way would have been open for the first bacteria to appear.

All this can be summarized by emphasizing that Prigogine's 'dissipative structures' that are needed for life can be developed from scratch, with no previous information needed, by using Eigen's hypercycles to set in motion the evolution of species and the survival of the fittest, at an extraordinarily simple chemical level.

Thus, genetic information is coded from the beginning by the survival of the most suitable chemical process, and it accumulates in small stages which are the *most probable* at each time point. The questions raised by those who feel intuitively that life is a phenomenon too complex to emerge simply by chance no longer stand up, and the quantitative statement of this problem, as expressed by Fred Hoyle and Chandra Wickramashinghe, has proved to be totally misleading. We can in fact readily accept that 'life' is a very probable physico-chemical phenomenon that will appear soon after the prerequisite conditions are met. On the Earth, it could easily have emerged in the time available after the biosphere emerged as a result of the cometary bombardment.

THE HISTORY OF LIFE

All living organisms develop from germs, that is to say, they
owe their origins to other living beings. But how did the
first living things arise?

A. I. Oparin, 1924 (quoted in *Origins of Life*,
ed. D. W. Deamer and G. R. Fleischacker 1994)

The primordial Earth

Four and a half billion years ago, the proto-Earth was completing its
formation. During the last accretion phases, its growing gravity had
increased the impact velocities, so that their energies had been transformed into more and more heat. Hence the proto-Earth became progressively covered with a thick layer of molten lava, possibly to a very
great depth.

The large-scale differentiation that separated the denser iron core
from the mantle of lighter silicates was triggered by this intense heat.
The last large impact occurred somewhat later, notably the one
which, by a tangential grazing collision, caused the appearance of a
transient ring around the Earth that rapidly became the Moon. The
smaller cometary impacts, however, persisted and ended by establishing, not only the atmosphere and the oceans, but also the minor
differentiation that separated the terrestrial crust from the underlying
mantle.

The chemical and isotopic evidence that the terrestrial crust formed
so early on was an enigma for geologists. It seems to be resolved by the
cometary bombardment, when chondritic silicates were plowed deeply
into the surface of the Earth after the separation of the core from the
mantle. The bombardment explains not only the cosmic abundance

ratios of the siderophile metals in the terrestrial crust, but also the progressive emergence of a less dense continental crust.

Four billion years ago, the cometary bombardment was 99% over. Its rate dropped quickly to a few comets per century, then per millenium. The still dense atmosphere had time to stabilize and slowly cool, above oceans which were at first very hot, then tepid.

At that time, the surface of the Earth was still covered with craters, scars of giant impacts comparable to those still visible on the Moon. The heat of these impacts caused rock differentiation. The crust of less dense rocks floated by isostasy (see Glossary) and thus rode higher than the average level of the oceans, now widely established through condensation of the large amounts of water vapor thrown off by the cometary impacts. Many craters submerged in the depths of the oceans. Continental islands were in the making, and thousands of circular lakes appeared because craters filled. Torrential rains resulted from the intense atmospheric turbulence caused by the heat of the impacts and the large amount of water vapor released.

The number of lagoons and shallows, alternately covered and uncovered by the ebb and flow of the seas, was very great, because the tides were gigantic. The Moon had begun to spiral away from the Earth, but in the epoch we are considering, it was still at least three times closer than at present. Its orbit was more inclined on the ecliptic than it is now, and in five days it completed its revolution around the Earth, which itself was rotating twice as fast as it does now. Since tidal force varies with the inverse cube of the distance from the Moon, the tidal amplitudes then were close to 30 times as large as at the present time.

In this epoch, the atmosphere induced a greenhouse effect large enough to hold the temperature of the oceans at tens of degrees warmer than at present, despite the weaker radiation of the Sun (which had just gone through a minimum about 30% less than the present level). This effect was due in part to a heavy stratospheric fog. This fog probably comprised the polymer of formaldehyde $(H_2CO)_n$ brought in by comets (H_2CO forms fine particles of polyoxymethylene). Later this greenhouse effect was promoted instead by large amounts of CO_2 and water vapor.

In spite of considerable losses of volatile matter caused by the large collisions, the atmospheric pressure would probably have been able to rise to 60 or 80 atmospheres, if a new phenomenon had not converted a large amount of the atmospheric CO_2 into solid carbonates. CO_2 is much more soluble in rain water when under high pressure. Hence, the rain became acidified with carbonic acid (H_2CO_3); it attacked the silicates in the rocks and transformed them into silica (SiO_2), with the consequent formation of solid carbonates (see p. 166), especially calcite $(CaCO_3)$ and dolomite $(CaMg(CO_3)_2)$, usually imprisoned in limestone. The temperature of the ground was now sufficiently low that carbonates were not destroyed by heat; they formed huge sedimentary rocks, and a large fraction of the atmospheric CO_2 was locked into the ground.

When carbonates are heated by volcanism (triggered by the collision of tectonic plates), they decompose and restore CO_2 to the atmosphere. The planet Venus is a good example of a situation where the very high ground temperature has kept all the available carbon as atmospheric CO_2. In contrast, planet Mars has remained cooler, hence most of the CO_2 from its primeval atmosphere (which has the same cometary origin as that of the Earth) should have been buried in big sedimentary layers of carbonates. This is a matter that future exploration of Mars must verify soon.

The epoch when the terrestrial crust became cool enough to stabilize the layers of carbonates remains uncertain; so it is not possible to specify the maximum atmospheric pressure reached on Earth. It is not impossible that it reached a transient peak of 30 or 40 atmospheres during the course of the first 500 million years, but it had to fall to much lower values as soon as the oceans cooled to 70 or 80 °C.

Because the acid rains attacked the silicates that made up nearly all the surface rocks, the process of trapping the CO_2 in the ground continued throughout the ensuing billion years. Soon, the partial pressure of atmospheric CO_2 was below the partial pressure of nitrogen; hence, nitrogen became the principal constituent of the atmosphere. Nevertheless, the greenhouse effect caused by water vapor and the residual CO_2 remained substantial and it more than compensated for the fainter brightness of the Sun during the first billions of years. There was an important consequence: the first oceans were first very hot and then very warm.

The geological evidence supports this interpretation of the history of the atmosphere. Sedimentary rocks 3.8 billion years old are known, particularly in southwest Greenland. In order to produce sedimentary rocks, acid rains (water plus CO_2) were needed to alter igneous rocks chemically. This implies that on Earth, the climate has always kept oceans in the liquid state, and also that a large amount of CO_2 was present in the atmosphere before being buried in large sedimentary layers.

Those small amounts of dolomite that are still found in ancient rocks are rich in compounds of *reduced* iron; this absence of oxidation reveals that in the distant past, there was no oxygen in the air. The more diversified sediments (3.4 billion years old) found in Swaziland and in the western part of South Africa, as well as the substantial dolomite sediments (2 billion years old) found in South Africa, confirm the continued presence of liquid water in the oceans and the progressive decline of atmospheric CO_2. This corresponded to the progressive emergence of more limestone and dolomite sediments throughout the geological ages.

So, for the initial 500 million years, the terrestrial atmosphere was in an intermediate state of oxidation–reduction, with no free oxygen and some hydrogen steadily escaping from the exosphere, but replenished by the steady supply of cometary gases and their molecular photo-dissociations by the solar ultraviolet. The large initial input of CO_2 and the small supply of nitrogen finally produced the early major constituents; but for a long time, the atmosphere contained reducing gases such as H_2, NH_3, CH_4 and CO, as minor constituents. These gases were eventually photodissociated, or oxidized, like CO into CO_2, by the very small amount of free oxygen resulting from the photo-dissociation of water molecules in the high atmosphere. Because of the steady escape of hydrogen from the upper layer of the atmosphere, the atmospheric gases became more and more oxidizing, but this was much later. It is reasonable to assume that conditions favorable for the emergence of life were present early on and lasted for a long period of time at many locations on the globe.

Life makes its appearance

Prebiotic molecules were probably introduced on Earth by the slow fall of cometary dust down through the atmosphere. I mentioned that this microscopic dust was spread out in immense quantities for hundreds of millions of years. When stopped by the surface of the primitive oceans, which were very hot, the dust presumably became enclosed in tiny foam bubbles that often form on the ocean surface. Alternatively, some dust sank into shallows and reached a clay base. Clay contains grains of montmorillonite that may represent the first favorable site for chemical evolution. However, the microscopic foam bubbles also suggest a possible origin for the first cell wall, bringing to mind Oparin's ideas (not to mention the romantic ideal of life emerging out of the sea foam, like Botticelli's Venus!).

It is not difficult to envisage, reproduced in all the oceans in billions of copies and over millions of years, some infinitely varied conditions where prebiotic molecules polymerized and induced the emergence of competitive hypercycles (see Chapter 5). Some amino acids formed short chains of peptides; some nucleic acids joined them in catalytic loops. Progressively, hypercycles selected those catalytic reactions best suited to link DNA fragments and peptide fragments. There was no random selection here, as Fred Hoyle would have liked us to imagine; instead, there is selection of the fittest by retroactive loops.

As de Duve suggested, phosphorylations resulting from the forma-tion of thioesters may have provided the phosphates. Such a hypothesis would explain the source of energy required to start the process. We know very little, but we can start imagining scenarios in which small fragments of RNA folded like hairpins. The two branches of the hairpin were mutually stabilized by complementary nuclear bases, giving rise to the first transfer RNAs (see Figure 5.5).

The future 'keys' to the translation of the genetic code would be the first to appear, in a simplified primitive form. For example, as sug-gested by the redundancies and synonyms in the actual genetic code, 12 or 16 nucleic bases could have assembled in such a way as to encode an amino acid in two 'letters' (instead of the present three).

From these different pieces of transfer RNA the retroactive loops of the hypercycles selected, from all possible combinations, the fittest for survival and thus produced the first 'protobionts'; that is, organisms the size of viruses, but which could multiply in their existing environ-ment and evolve by mutation.

We still do not know many details, because the smallest known virus comprises a DNA loop that contains about 10 000 nucleotides. The present viruses act parasitically since they depend on the prior existence of the host that they infect. It is possible that they have undergone a regressive evolution. Whatever their origin, they illustrate how a simple primitive organism of their size was able to self-organize from prebiotic molecules in order to reproduce.

Originally, localized sources of energy had to be used, like those of the warm sulfurous springs, which could also supply sulfur for thioesters; inorganic phosphates could be used as the source of the organic polyphosphates. However, the use of solar light for photosynthesis was presumably an important step toward independence from the local sources of energy; hence we understand the existence of an evolutionary thrust toward solar light. With such a driving force, photosynthetic bacteria would have emerged fast.

The origin of life remains an immense problem and the gaps in our knowledge are still countless. For 20 years, progress in molecular biology has been spectacular, but there is still much to be done. The fossil record that is still present in the genomes of all living species might soon be used to considerably expand our understanding of the origin of life.

The world of bacteria

For about 3 billion years, the Earth was the dwelling of only a single kind of life – bacteria – which developed and multiplied everywhere there was water. But where there was no water, the continents remained sterile and desert-like. Marine multicellular animals and aquatic plants emerged only 600 million years ago. Land plants began to overrun the continents only 420 million years ago, but we shall discuss this later.

I have already mentioned that the bacteria can be divided into three groups (Figure 5.10). These are the Archeobacteria, the true bacteria and the Urkaryotes. A large number of true bacteria are known. The true bacteria are often called the Eubacteria, in order to distinguish them from the two other groups. Conversely, very little is known of the Archeobacteria, but it is a very old group and it is suspected that a large number of their species are now extinct, eliminated by natural selection. This is suggested by the diversity of their RNA sequences, and by the separate ecological 'niches', with conditions reminiscent of those on the primitive Earth, where the extant species have survived.

In particular, the thermoacidophils are an example of Archeobacteria that must be very close to the universal common ancestor, because they are found only in very hot sulfurous springs that correspond well to our description of conditions favorable for the emergence of life on primitive Earth. Another example is the methanobacterium which cannot use sugars, proteins or any carbohydrates as food. These bacteria have a totally unexpected way to produce the energy that they need, and this can be summarized in the following equation:

$$CO_2 + 4H_2 \rightarrow CH_4 + 2H_2O + energy$$

The process is based on the reduction of carbon dioxide (CO_2) and oxidation of hydrogen (H_2) (which the bacteria have difficulty finding nowadays), in order to free the energy they need, while giving off methane (CH_4). Since the oxygen present in the air was poisonous to them, some found shelter in the stagnant water of marshes, while other species have survived only in the digestive tract of ruminants; this is true symbiosis because they help the cow, for example, to digest the cellulose present in hay and grass. These bacteria must have appeared at a time when there was no oxygen, but when the atmosphere of the primitive Earth still contained some hydrogen.

Let us now say a few words about Eubacteria (the 'true bacteria'). By comparison of their different common features, they can be traced back to their ancestors. Most but not all Eubacteria are photosynthetic; that is, they use solar light as their only energy source. Those which are no longer photosynthetic have obviously lost this feature in the course of their evolution. This is the case for a common bacterium of the human intestine, Escherichia coli. The genetic instructions written in its DNA are very similar to those of a group of purple bacteria that are still photosynthetic. Photosynthesis is a feature which appeared very early in evolution; we have already mentioned that the earliest known fossil bacteria (Figure 5.3.) were already photosynthetic.

We come now to the third ancestral group of bacteria, the Urkaryotes. This is the most important group for us, because it gave rise to eukaryotic cells, which were to form animals and plants. The urkaryotic bacteria do not seem to have survived in their primitive form, because they were 'infected' by eubacterial parasites that sheltered in their cell membrane. Those surviving have eventually benefited greatly from the parasites' presence, which helped them to survive. It is a classic case of 'symbiosis', which we shall discuss later in the context of eukaryotic cells.

Needless to say we have not been able to reconstruct all the details in the evolution of the three divisions of primitive bacteria during the first 2 billion years of their evolution. The key to our ignorance probably lies in the unimaginable diversity of species that are now extinct. In addition to the most ancient fossil bacteria (recovered in Australia and dating from about 3.5 billion years), the only abundant fossil identification is that of colonies of blue-green algae (now termed cyanobacteria), which have formed stromatolites in large numbers and in many places. In contrast to the green algae, the blue-green monocellular algae are much less of a plant than a bacterium (i.e. without a cell nucleus); this is the reason why they are more correctly called cyanobacteria.

Until the emergence of these organisms, the oxygen content of the air was limited to about 0.1% of its present value. This slight trace of oxygen was produced by the photodissociation of water vapor by solar ultraviolet. The oxygen so produced absorbed the ultraviolet wavelengths which would have been able to continue the photodissociation of water, and the production of oxygen from water stopped at the 0.1% level. But another source of oxygen had appeared: throughout more than 2 billion years, the cyanobacteria present in the stromatolites slowly poured increasing amounts of oxygen into the atmosphere.

Cyanobacteria use solar light to convert carbon dioxide into a nutrient (a sugar). In darkness, they consume some of this nutrient by 'burning' it slowly with oxygen, but the oxygen excess formed remains toxic to them so they release it into the air. This is the reason why the oxygen content of the air began to climb steadily, and increasingly protected the continents from the sterilizing effects of ultraviolet. At low tide, stromatolites in the shallows protect themselves from the harmful ultraviolet by layers of redundant limestone; other bacteria prefer a thick layer of sea water. But the situation changed slowly and, with enough oxygen to make a protective layer of ozone, new forms of life dared to leave the sea.

The eukaryotic cell

At an undetermined moment, between about 1.5 and 2 billion years ago, the eukaryotic cell emerged, opening up a new and highly significant fork in the chain of life forms. In contrast to the prokaryotic cells of bacteria, eukaryotic cells are 3 to 10 times larger in size, therefore 30 to 1000 times bulkier. They protect their chromosomes, made of DNA and proteins, within a well-defined '*nucleus*' enclosed in a membrane, while the prokaryotes have only a circular strand of DNA which floats freely in the cell (see Figure 6.1).

The eukaryotes shelter '*mitochondria*', which provide energy from the oxidation of simple organic compounds; the eukaryotes of green algae also contain '*chloroplasts*', organelles whose function it is to absorb solar light by means of a pigment, and convert CO_2 into nutrients by photosynthesis (as plants do).

Mitochondria and chloroplasts seem to derive from bacterial parasites of the ancestral Urkaryotes. They have formed a symbiotic relationship with their host, both finding in it an advantage that improves their chances of survival. Not only do these organelles of the eukaryotic cell have a genetic system separated from that of their cell host, with separate DNA and transcription and translation processes, but they also possess some autonomy.

It is probably not a coincidence that the first monocellular eukaryotes (the *protists*) appeared at the time when free oxygen began to grow in the atmosphere about 2 billion years ago. Practically all eukaryotes need oxygen to live, whereas a large number of bacteria are killed by oxygen; however, some have developed strategies to survive in an oxidizing atmosphere.

Protists are single eukaryotic cells, no longer bacteria, but still microbes in size. They constitute the common ancestors of animals, plants and fungi. Among the protists, the protophytes are the ancestors of plants (they already possess a photosynthetic pigment), while the protozoans are the ancestors of animals. The most important event in the evolution of the protist was probably the invention of sexual reproduction, because it allowed a spectacular acceleration in the rate of evolution. This new function had to come after '*mitosis*', division of the cell into two equal parts where all its elements are reproduced

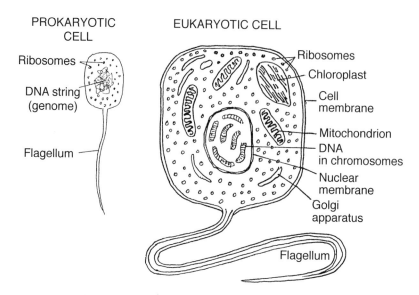

PROKARYOTIC CELL

Ribosomes

DNA string (genome)

Flagellum

EUKARYOTIC CELL

Ribosomes

Chloroplast

Cell membrane

Mitochondrion

DNA in chromosomes

Nuclear membrane

Golgi apparatus

Flagellum

FIGURE 6.1   The prokaryotic cell (that of the bacteria) is contrasted with the eukaryotic cell. Bacteria are prokaryotes that appeared on Earth about 3.8 billion years ago; the monocellular eukaryotes appeared at the earliest 1.9 billion years ago; that is, about 2 billion years later than the bacteria. The principal feature of the eukaryotes is that their DNA is enclosed within a membrane, the *nucleus* of the cell. Moreover, it is often (not always) wrapped around a bundle of proteins; this forms one or several '*chromosomes*'. The *Golgi apparatus* store the metabolic by-products. *Mitochondria* carry out cellular respiration; *chloroplasts* absorb light to make food by photosynthesis. These three types of *organelle* seem to come from ancient bacteria that have given up their independent life to enter into symbiosis with the eukaryotic cell. The numerous *ribosomes* carry out the assemblage of proteins, translating fragments of the messenger RNA which leave the nucleus after having transcribed part of the DNA. Not all different species possess each organelle described. Notably, the flagellum is absent from multicellular eukaryotes.

equitably. These two events (mitosis and sexual reproduction) must have occurred between 1.7 and 1.5 billion years ago.

But this is about 3 billion years since the formation of the Earth – how had the planet been evolving during this long period?

The Earth and the Moon

The Earth has not always rotated about its axis once in 24 hours. Statistically, the accretion process led to planets that rotated in 10 hours; but each of the last collisions introduced a random perturbation that in some cases were large, particularly when the impacts were nearly tangential and when the colliding objects were already very massive. This is the origin of the arbitrary orientation of the axes of different planets, which are often far from perpendicular to the plane of the ecliptic. The formation of the metallic core of the Earth reduced its moment of inertia, thereby accelerating its rotation. In addition, there was the tangential collision with the proto-planet that formed the Moon, which again accelerated the rotation of the Earth.

But when the Moon formed, it induced gigantic tides in the Earth's oceans, as well as in its crust and mantle; the tides began strongly to retard the rotation of the Earth, so that its period of rotation lengthened – very quickly at first, then more and more slowly because the Moon's orbit was widening. This process, which slowly dissipates the gravitational energy of the Earth–Moon system, is well understood and its actual rate has been verified by observations. In particular, the distance from the Earth to the Moon has been measured by light from a laser reflected by mirrors left on the Moon by the astronauts. These measurements reveal that our satellite is separating from us, increasing its average distance by 4 cm per year. Extrapolating back in time, this rate would bring the Moon and the Earth together a bit too early. It implies that the dissipation of tidal energy was somewhat less in the past; this is probably due to the changing shape of the continents.

When it widened, the Moon's orbit also became very flattened onto the ecliptic, so that it is now 5° away from it. The mass of the Moon exceeds 1.2% of that of the Earth, which is large for a satellite. This mass has a stabilizing effect on the motions of the Earth. As a result of its rotation, the Earth had its poles flattened and developed an equatorial bulge. The Moon attracts it, causing precession and nutation of the Earth's axis, which behaves as a spinning top.

Precession causes the Earth's axis to swing around every 26 000 years, while holding its inclination constant at 23.5° off the normal to the ecliptic. Nutation is a slight rocking of the Earth's axis over 18.6 years around its average inclination. Combined with the

quasi-periodic variation of the Earth's orbit due to planetary perturbations (mostly from Mars), these motions influence the distribution of the climatic regions on the Earth, and in particular explain the
recent glaciations. However, in remote times, in spite of the weaker
radiation of the early Sun, the greenhouse effect of much larger
amounts of CO_2 was strong enough to completely remove the possib-
lity of any glaciation. Let us first examine in detail how the atmosphere
has evolved.

The evolution of the atmosphere

The rate of decrease of atmospheric CO_2 can in principle be evaluated
from the mass of sedimentary carbonates at different geological
epochs, because the reaction is well known quantitatively. By diluting
CO_2 in water (H_2O), the rain forms carbonic acid (H_2CO_3) which
transforms the silicates ($-SiO_4$) of the rocks into carbonates ($-CO_3$):

$$2H_2CO_3 + Ca_2SiO_4 \rightarrow SiO_2 + 2H_2O + 2CaCO_3 \text{ (calcite)}$$
$$2H_2CO_3 + MgCaSiO_4 \rightarrow SiO_2 + 2H_2O + MgCa(CO_3)_2 \text{ (dolomite)}$$

The warmer the acid water, the faster the reaction. The total of known
carbonates corresponds to more than 30 atmospheres of CO_2, but the
most ancient sediments are difficult to find. By the subduction, or
'pulling down', of the continental crusts, there has been a steady re-
cycling of all sediments, and the most ancient sediments may hold at
the present time a substantial fraction which is hidden too deep to be
found.

There is little doubt, however, that the intense hot rains of the early
atmosphere produced a large amount of carbonate sediments and a
rapid drop of the CO_2 pressure. Even when the partial pressure of CO_2
reached a very low level, it continued to induce, with water vapor, a significant greenhouse effect that was large enough to prevent the formation of polar caps and the appearance of glacial periods throughout
the first 2 billion years. But this could not last. During the last 2 billion
years, ice caps emerged at the two poles and, periodically, they became
vast, extending sometimes to normally warm latitudes. We will discuss
these 'glaciations' later.

Before going any further, let us discuss the 'greenhouse effect' in
depth. The greenhouse effect takes its name from the garden or botan-

ical greenhouse. Its glass panes protect vegetables or flowers from cold, while allowing sunlight to enter. The panes hold radiant heat within the greenhouse because they are opaque to the infrared. This stems from the radiation laws: the higher the temperature of the radiating body, the shorter the wavelengths where the radiation concentrates. Because the Sun's surface is close to 6000 K, its radiant energy is concentrated in the visible light, say, approximately 0.5 μm. The soil and plants of the greenhouse, on the other hand, are close to 300 K. Since this temperature is 20 times less than that of the Sun (measured in kelvins, from absolute zero), the energy radiated by the soil inside the greenhouse is concentrated at wavelengths 20 times longer; that is, in the infrared from 5 to 20 μm, with a maximum near 10 μm.

The greenhouse effect of CO_2 is due to its transparency to visible light and its opacity in the infrared; more specifically, to the existence of absorption bands in the CO_2 spectrum, between 3 and 30 μm. In this region of the spectrum, CO_2 has two very strong absorption bands, one centered on 4.3 μm and extending from 4.1 to 4.5 μm, the other centered on 15 μm, extending from 13 to 17 μm. There is also a weaker band from 9 to 10 μm. It is because these bands absorb the infrared energy radiated from the ground that the CO_2 molecule heats up and warms the atmosphere, which then radiates this heat back toward the ground.

The limit for the greenhouse effect is reached when the ground is so warm that it begins to radiate at longer and longer wavelengths that can escape between 6 and 9 μm – a 'window' left transparent by all the gases present in the atmosphere (mostly CO_2 but also water vapor). It is difficult to compute the Earth's temperature caused by the greenhouse effect some billions years ago, because of too many poorly known variables (cloud cover, traces of other opaque gases, the albedo of the Earth at that time, the temperature distribution in the upper atmosphere, etc.)

However, it is probable that the window of transparency left open by H_2O and CO_2 between 6 and 9 μm stabilized the ground temperature at a value some tens of degrees higher than at the present time. This stabilization lasted for a very long time, until the atmospheric CO_2 fell to a very low level and the oxygen rose in proportion. This is a good point to consider in detail the process by which CO_2 is transformed into oxygen (O_2). I have already mentioned that photosynthetic bacteria use solar light to convert atmospheric CO_2 into nutrients. Photosynthesis is a very complicated chemical process using many steps, but it can be summarized by an equivalent global reaction:

$$nCO_2 + nH_2O + \text{light} \rightarrow (CH_2O)_n + nO_2 \text{ (freed)}$$

$(CH_2O)_n$ stands for the general formula of the carbohydrates, such as the sugars, where the subscript n may take many values. These nutrients are later partially used to supply the energy needed for biological functions, by an oxidation process (the equivalent of a slow combustion) that is the inverse of photosynthesis:

$$(CH_2O)_n + nO_2 \rightarrow nCO_2 + nH_2O + \text{energy}$$

Only a small fraction of the carbohydrates are used to produce energy; hence CO_2 is never completely restored. The rest of the carbohydrates are used to build the cell of the bacterium; upon the death of the organism, carbon goes into the soil whereas the excess oxygen remains in the air.

This free oxygen first oxidizes everything possible on the ground; only thereafter does the remaining excess become transiently available in the atmosphere, where it stays far from thermochemical equilibrium, since it is always available for slow oxidations or fast combustions.

To summarize the two steps that transformed the early atmospheric CO_2: the first was the formation of sedimentary carbonate rocks that trapped large amounts of CO_2 in the ground; the second was taken by bacteria which removed only carbon and released a steady supply of oxygen into the atmosphere.

The age of the stromatolites

Stromatolites are limestone deposits with layered features which shelter colonies of cyanobacteria (formerly called blue-green algae). The stromatolites steadily multiplied from 2.4 to 2 billion years ago, as the cyanobacteria converted more and more CO_2 into oxygen. There are telltale signs that atmospheric oxygen continued to be insignificant (about 0.1%) up to 2.4 billion years ago. Then the amount began to rise rapidly and reached 3% some 2.0 billion years ago (growing by a factor of 30 in 300 million years!).

At that time, there must have been an explosion in the number of stromatolites, which is confirmed by their fossil remains found the world over. Stromatolites had been in existence for eons; they were very similar to the first fossil remains of the photosynthetic bacteria discovered by Schopf and dating back to 3.8 billion years ago. The sudden increase in the oxygen content of the air, beginning 2.4 billion years ago, indicates perhaps that there was nothing left to oxidize on the ground. The oxidation of reduced mineral deposits and the oxygenation of the oceans was a formidable task that the stromatolites must have already accomplished by 2.4 billion years ago.

Figure 6.2 shows graphically how the oxygen content of the air increased dramatically 2 million years ago. Previously, the danger of exposure to the solar ultraviolet confined bacteria during the daytime to a depth of 10 m below sea-level, or between the limestone layers of stromatolites. Later, the increasing oxygen level established a layer of ozone in the upper atmosphere. Ozone is the triatomic form of molecular oxygen (O_3), produced by the ultraviolet photodissociation of O_2; O_2 then splits into two atoms that immediately recombine into ozone, by collision with another O_2 molecule:

$$O_2 + \text{ultraviolet light} = 2O \quad \text{and} \quad O_2 + O = O_3$$

Unstable as it is, the ozone molecule is easily destroyed by collisions with other molecules, but the protection afforded by even a very thin layer of ozone is sufficient to prevent the sterilization of bacteria, because of its ability to strongly absorb ultraviolet light of 0.2 to 0.3 μm wavelength.

Bacteria took such a long time to leave the water and diversify, because during this period they had no protection from the lethal ultraviolet light. But as soon as the oxygen content of the atmosphere reached 3%, there was enough dissolved oxygen to render water poisonous to many bacteria. This triggered an extraordinary revolution in the life forms: small bacteria took refuge as parasites inside larger bacteria. Those which survived soon came to live in symbiosis with their hosts, providing welcome functions such as respiration (providing the dreaded oxygen to digest food) or photosynthesis (using light to synthesize food). The host became larger and larger; it protected its genetic recipe inside a membrane, and partitioned the cell in different ways to preserve order. Finally, it became a new microbe, too

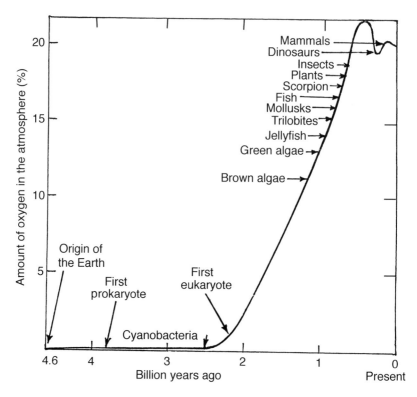

FIGURE 6.2 The oxygen content in the atmosphere as a percentage of atmospheric pressure. During the first 2 billion years, traces of oxygen came from the photodissociation of water vapor. Soon after the arrival of the first prokaryotic cells, the cyanobacteria transformed CO_2 into oxygen by using solar light; 3 billion years ago, they had colonized the world, but there was an explosion in their number 2.5 billion years ago (small arrow), which triggered a rapid growth in the oxygen content of the air. As soon as the amount of oxygen was large enough to absorb the lethal far ultraviolet (O_2) and near ultraviolet (O_3 layer in the upper atmosphere) radiations, life could emerge from the oceans; algae and then plants took over oxygen production and almost exhausted the CO_2, triggering an instability in the oxygen content that followed the giant fern forests of the Carboniferous era.

big to masquerade as a bacterium. The first eukaryotic cell was born: it was a protist.

The much more complex eukaryotic cell allowed more diversification; this soon led to two kinds of protist. The first kind was the Protophytes: they incorporated *chloroplasts*, former bacteria which contained the right pigments to absorb either visible, or infrared, or ultraviolet light (0.3 to 0.4 μm, the non-lethal wavelengths that ozone does not stop). These pigments allow photosynthesis to occur and are already very similar to the chlorophyll found in future plants.

The second kind of protist was the Protozoa; they incorporated *mitochondria*, which made use of the new availability of oxygen to produce energy by the slow combustion of nutrients.

This also led to the formation of the first colonies. The bacteria were already joining together to form filaments (see Figure 5.3) or layers like the cyanobacteria, protected in stromatolites by limestone layers (Figure 6.3). The protists soon became complex enough to specialize and distribute tasks within a colony. Colonies where individual cells had specific tasks of a different nature soon become too interdependent to be separated; they become a multicellular animal. This happened to algae that lived in the oceanic plankton. The algae were very diversified. Some had mitochondria and foreshadowed the future animals, but most were photosynthetic and wanted the best pigment for their chloroplasts; some were brown, purple, gilt-coloured or green. The last of those had already found chlorophyll, the fundamental molecule for plant evolution.

But evolution is inventive. Diatoms are algae which have developed a siliceous bivalve shell for protection. Large beds of giant kelp develop near coasts and are laid down on beaches; more than 1000 species of kelp have been described, and all are photosynthetic.

But there are too many species, genera, families, orders, classes, phyla . . . More than 1000 species of plankton have been catalogued. Some of them invented sexual reproduction, which accelerated diversification in a spectacular way. There were: the Radiolaria, with a fine siliceous skeleton in grid form; the Foraminifera, encased in a porous shell reinforced with calcium carbonate; and soon tiny Crustacea which were the ancestors of the future insects.

FIGURE 6.3 Stromatolites: these are colonies of cyanobacteria, often called 'blue-green algae', because they look superficially similar to other algae (green, brown, etc.). They are actually bacteria (with a prokaryotic cell), whereas other algae are aquatic plants composed of eukaryotic cells. The cyanobacteria are probably the longest living species on Earth. They have survived to the present by taking refuge in the salty waters of tropical lagoons that plants and animals avoid. The photo shows living stromatolites in a salty lagoon at Shark Bay, on the tropical west coast of Australia. Each generation uses a mat of limestone waste product to protect itself from the ultraviolet. The superposition of many generations takes the form of a rounded rock, which is what is called a stromatolite. The cyanobacteria are photosynthetic bacteria; their oxygen production seems to have been sufficient to develop the first oxygen and ozone protecting layer, which was going to allow life to come out of the oceans and invade the continents.

Marine animals

Some of the protists that we have just mentioned, the Foraminifera, expanded their single cell to several centimeters in diameter. They could no longer be called microbes, although they were still unicellular. But there is a limit to the size of a single cell, because the internal connections become slow and the chemical processes lose their efficiency.

A very large size is an undeniable advantage for defense against predators, but the same result can be achieved more easily by grouping a large number of cells into a single colony. The case of the *volvox* is typical. Although capable of swimming separately, the cells of the volvox usually gather and form a small hollow sphere 1 mm in diameter; the numerous cells synchronize the beating of their flagellum to move the sphere as if it were a single entity.

The next evolutionary stage was probably achieved 1 billion years ago; sponges appeared, followed by corals and jellyfish. Sponges and corals are colonies of unicellular organisms that build their own protection: a soft and flexible mass for the sponges, hard and chalky for the coral polyps. The jellyfish, or more generally the coelenterates, are organisms that eat plankton and have an axis of symmetry. They are basically the simplest of all true animals, because they possess distinct organs: their bodies have a single cavity connected to the outside by a 'mouth', which also serves to excrete digested food (all coelenterates are carnivores).

In 1947, very delicate fossil traces of coelenterates were found in the sandstone of the Ediacara Hills, in South Australia. They date from nearly 600 million years ago. Except for the fossils of ancient bacteria, and the numerous stromatolites, these are the most ancient fossils of non-microscopic animals that have been found up to now. Subsequently they have been found in many other places in the world.

A little less than 600 million years ago, numerous fossils appear nearly everywhere in the sedimentary limestones. They are typically brachiopods, crinoids and mostly trilobites. They are easy to identify because their limestone shells have survived. These shells are the first exterior skeletons (exoskeletons), designed to protect the mollusks.

Glaciations

The atmospheric greenhouse effect remained intense as long as there were large amounts of CO_2 in the air. However, the amount of CO_2 steadily diminished; this resulted in a slow cooling of the atmosphere in spite of the very slow rise in solar radiation. At 2.4 billion years ago, there was probably still about 20% CO_2 when the population explosion of the cyanobacteria produced a rapid depletion. The evidence for this is that, in a matter of only 1.7 billion years, the oxygen content of the atmosphere rose to 20%.

Some scientists claimed that before the oxygen increase there had been a large, global glaciation about 2.3 billion years ago. It was well-documented in geology, particularly with glacial remains in Canada, Finland, India and Siberia, as far as the southern region of Lake Baikal. However, the interpretation of these glacial remains has been strongly contested very recently, and it is more logical to believe that this 'first glaciation' never existed. According to Figure 6.2, a glaciation some 2.3 billion years ago does not make sense because there was still about 20% CO_2 in the atmosphere that would be used by cyanobacteria, and later by algae and plants, to make the 20% oxygen. However, we will bow to tradition, and call 'second glaciation' the one that lasted from 900 to 800 million years ago and coincided with the total depletion of the CO_2 and the maximum amount of O_2 in the atmosphere. A 'third' glaciation occurred between 700 and 650 million years ago. The harsh conditions introduced by these last two glaciations, which brought icy polar caps down to mid-latitudes, have certainly played a role in the diversification of species and their struggle for survival.

With the greenhouse effect considerably reduced because CO_2 was now well below the 1% level, the different movements of the Earth were going to have noticeable effects on the climate. These secular motions of the Earth are its precession, its nutation and the change of the eccentricity of its orbit due to perturbations by the other planets. Another important effect on the climate of different regions is continental drift. Although the axis of rotation of the Earth remains attached to its internal mass, the thin continental plates float on the 'asthenosphere'. The asthenosphere is the part of the mantle which remains fluid because of its high temperature. Thus, the continental plates may remain in constant motion with respect to one another, as well as relative to the axis of rotation of the Earth.

Because the continental plates often meet, their collisions unleash large amounts of energy, part of which is transformed into heat and produces volcanic activity. The heating of sedimentary carbonates causes their decomposition, and the volcanoes return CO_2 to the atmosphere. The epochs of major volcanism brought CO_2 back into the air, thus increasing the greenhouse effect again.

Several times in the past, either the upheaval of the continental plates or the upsetting of the climate caused mass extinctions of living species. These extinctions are well documented in the fossils of each geological layer, at least where these layers can be easily studied, i.e. for the last 600 million years. The traditional boundaries of the big geological periods have been set by the dates of those major extinctions which changed the nature of the observed fossils.

The largest of the mass extinctions occurred near the end of the Cambrian (515 million years ago), then of the Ordovician (439 million years ago), of the Devonian (367 million years ago), of the Permian (245 million years ago), and of the Cretaceous (65 million years ago). Finally, in the Quaternary, there was a large extinction which is still going on, and began only 50 000 years ago.

With the exception of those of the Permian and Quaternary, the other mass extinctions have been associated with a warming of the climate, the increase of the greenhouse effect being attributed to more volcanic activity due to collisions of the tectonic plates. A global temperature increase of several degrees is sufficient to endanger many tropical species. The fossil remnants of the tropical marine fauna and flora show how much they were devastated at the end of the Cambrian, the Ordovician and the Devonian. The big catastrophe that occurred at the end of the Cretaceous will be discussed below.

There is no space here to examine the evolution of all forms of life, but before leaving the oceans of the Cambrian, the mollusks and the arthropods must at least be mentioned briefly, if only because of the immense variety of their species, which began to appear in the sea more than 500 million years ago. More than 110 000 species of mollusk and more than 1 million species of arthropod have been catalogued. Mollusks have a soft body protected by a shell of calcium carbonate, while arthropods have a body and legs formed in distinct segments. The insects are arthropods that have *three* pairs of legs; a half million species of them are known.

Sexual reproduction

In the evolved forms of life, the genetic message carried in DNA had lengthened considerably. For the eukaryotic cell, it became so long that it separated into several pieces, each packed around those proteins used for its assembly and replication. These pieces are the '*chromo-somes*' (see Figure 6.1); they also contain *introns* (long unused segments of the genetic message, often an incoherent gibberish that is cut off before transcription by the messenger RNA, but which copying errors can reactivate).

In multicellular organisms, the embryo starts out from a single cell that divides into 2, then 4, 8, 16, 32, etc. But the neighboring cells are mutually influenced by enzymes that make them specialize, in order ultimately to build the body of a plant or of an animal, by a process that is only beginning to be understood. Basically, enzymes allow the DNA to express part of its message, and suppress the rest. Different enzymes develop muscles, or nerves, or skin, or teeth, or nails, or blood.

The mutations that lead to all this complexity must also have produced an immense number of errors, of which most were quickly eliminated by competition and survival of the fittest. The huge number of branchings in the complexity tree explains the sudden profusion of multicellular life forms. One of those errors led to the existence of a new type of cell, used only for reproduction. Some bacteria had already invented a nearly similar process, sexually mixing their DNA, as was discovered by the American biochemist Joshua Lederberg (1958 Nobel laureate for medicine); but only eukaryotes were able to develop sexual reproduction with all its consequences.

The ordinary non-sexual cells (called the *somatic* cells) contain an even number of paired chromosomes that are nearly, but not quite identical. In order to reproduce sexually, a cell divides its double series of chromosomes into two simple series (a process called '*meiosis*') that lead to two sexual cells, each having *half* the original number of chromosomes.

In order to produce an offspring, two sexual cells of different types must combine. At first glance, this would seem a needless complication. However, meiosis achieves a double goal. First, the offspring inherit all the common permanent properties of the species; secondly, they inherit a mix of the variations that come from the differences between the two parents, resulting in a greater genetic diversity in the next generation and a larger variation of the species. This is good for the struggle for life of the species, in particular when the environment is changing (emergence of oxygen in the air, climatic change because of volcanic activity, glaciations, etc.).

To make this process more effective, evolution introduced a asymmetry between the two sexual cells. By and large, the male cell has a flagellum (see Figure 6.1), allowing rapid motion in competition with other male cells in the search for a female cell, which conversely merely waits for fertilization. The male cell brings only half of the chromosomes, while the female cell, which is larger, also provides the nutrients for early growth, as well as some organelles for the completed new cell. *Mitosis* (the duplication of the cell with the usual doubling of the chromosomes) will then begin the embryo's development.

As soon as sexual reproduction appeared, the consequent greater variability of the species helped their rapid adjustment and survival under changing conditions. This adaptability allowed the marine animals to take an extremely difficult evolutionary leap, namely to leave their aquatic environment for the first time. Taking advantage of the recent emergence of free oxygen in the atmosphere, they came out of the oceans and invaded the continents.

Life invades the continents

The flux of solar radiation had increased steadily for 4 billion years. About 500 million years ago, although still 3% weaker than at the present time, it had become strong enough to prevent the occurrence of long glacial periods, even though there was very little CO_2 in the air. Only short glaciations appeared, extending the polar caps considerably but for short time spans. Their size fluctuated because of the secular movements of the Earth's axis, as well as because of the changes in eccentricity in the Earth's orbit around the Sun.

Some 515 million years ago, the massive extinctions at the end of the Cambrian left room for the expansion of a large number of new species. Starfish appeared; mollusks hid in shells of calcium carbonate. Arthropods such as trilobites soon produced crustaceans enclosed in a shell of chitin (a polysaccharide). But a great event was in store, because the ancestor of the fish had just invented the spinal column (while still keeping outside protective plates). It possessed gills, rudimentary lungs that filter the water to remove the dissolved oxygen.

Bones were a new material, necessary later to carry the body weight when animals left the water. Nearly as solid, but more flexible than cast iron, bone is cartilage mineralized with phosphate and calcium carbonate. The cartilage is formed of fibers of collagen, a protein whose molecule is made up of a thousand amino acid building blocks. Bones hold 88% of the phosphorus and 99% of the calcium in our bodies.

Like the modern lamprey, the ancestors of the fishes did not yet have bones in their jaws. The sharks then appeared and decimated the first fish during a glaciation of the southern hemisphere 430 million years ago. This glaciation destroyed half of the marine fauna; however, new species of fish appeared. Glaciations had already changed their nature at that time, and depended mostly on the secular motions of the Earth. Fluctuations of the now very low content of CO_2 set off a sequence of glacial and interglacial ages, during which the polar caps and the mountain glaciers advanced and receded, making the average sea-level fall and rise again.

At 410 million years ago, the continental plates that now form Europe and America collided. The average sea-level had fallen by 200 m, draining large reaches of the continental margins that had been under water. Numerous species of green algae progressively adapted to survival in dryer and dryer environments. They were the first plants – without roots, without leaves and without a stem. The earliest of these species have now disappeared, because they evolved quickly toward marshy plants (like the peat mosses), not completely adapted to living without water. Most survived, however, during long periods of drought.

After the peat mosses, the lycopods and horsetails appeared in marshes followed by ferns, then cycads and ginkgoes which foreshadowed the conifers; 390 million years ago, the explosion of vegetation on the continents (which had been sterile up to then) became extravagant. Tree-like ferns reached a height of 12 m and grew in large forests.

Now that there were plants on the land, they could serve as food, and it became possible for animals to leave the water and invade the continents. The first animals capable of breathing air were the scorpions. Their gills adapted to a new kind of breathing. Other species of arthropods soon followed, and the forests of tree-like ferns were quickly invaded by a large variety of insects.

The continents were now covered everywhere by dense forests that consisted of giant ferns and cycads both seeking light. The competition to reach light was intense, because it is needed to turn on photosynthesis that feeds plants by fixing the atmospheric CO_2. This competition led to giant vegetation that superficially resembled our present equatorial rain forests. It is hard to imagine marshy horsetails that grew trunks 30 m high and 2 m in diameter, but we find their fossil remnants in coal. These enormous forests depleted the CO_2 content of the air until it reached extremely low values.

About 360 million years ago some fish learned how to use their fins as legs, probably to drag themselves into the swamps, in search of deeper water, during periods of drought. They also learned how to use their gills as lungs. This 'apprenticeship' is of course a figure of speech, since it still comes down to random evolutionary features, and survival of the fittest. Soon the class Amphibia appeared. These are semi-aquatic vertebrates whose larvae live in water. At some point, through metamorphosis, they become adults capable of living on dry land, but they always like to return to water. All the amphibians – toad, frog, triton, salamander, etc. – return to water to lay their eggs; these eggs remain fragile because they do not have a protective shell.

Another variety of vertebrates soon developed – Reptilia. These are the snakes, turtles, lizards, crocodiles, etc. Since there was not yet any predator on dry land, they discovered the advantage of laying their eggs out of the water. To protect the egg, they enclosed its semipermeable membrane with a shell of calcium carbonate. The egg consists of a single female sexual cell, swelled by nutrients for the embryo, which must be fertilized by a male sexual cell to develop into an adult.

Even though many reptiles remained amphibious, they followed the example of the amphibians and began to invade the forests in search of food. Being carnivores, they hunted arthropods, such as scorpions, spiders, centipedes and mostly insects, which by then had multiplied greatly.

By 300 million years ago, some insects had found a clever way to escape from toads and lizards. Transforming their chitin shell into wings, the first dragonflies appeared, forerunners of the may bugs, bees and flies.

When the first wings spread from the body, their function was perhaps entirely different. First, it may have been a shield against the cold; later, extension of the wings in sunshine allowed the body to warm up quickly, by stimulating the blood flow. In the course of evolution, it is not unusual to see organs appear at first for a totally different purpose from that – unforeseen – which will become their final function.

Another example is the eye, which began first as a photosynthetic organ, like the chloroplasts, whose purpose was to use light as a source of energy, before progressively coming to be used as a source of information.

By 250 million years ago, some lizards had developed into animals of very large size, like the crocodiles. One group of these giant lizards developed the ability to run on its hind legs, and became agile and fast; these were the thecodonts. These animals soon engendered two quite different kinds of saurian or dinosaurs, some of which became truly gigantic. The dinosaurs dominated the Earth from 225 to 65 million years ago; that is, for a time span of about 160 million years.

With the thecodonts, we have reached the dawn of the age of the dinosaurs.

The age of the dinosaurs

Between 350 and 400 million years ago, several of the large continental plates collided and melded together. At first, this huge plate moved toward the south pole. Later, it began to move northward again; 300 million years ago, it ran into the continental mass we now call North America and Europe; shifting west, Siberia also ended by colliding with Europe, thereby forming one single gigantic continent that ran from one pole to the other.

This continent was called 'Pangaea' in 1912 by the German meteorologist Alfred Wegener. Although supported by factual evidence, his theory of continental drift has long been treated with scepticism. Only after the Second World War was it definitely proved and accepted, in particular thanks to the variable strips of magnetic fields locked in sediments in the bottom of the Atlantic Ocean, which are a record of the relative motion of the seabed with respect to the magnetic poles.

With artificial satellites, it has become possible to locate points on the Earth's surface with a remarkable accuracy. In particular, it is possible to measure within a few centimeters, the distance from a point in Paris to a point in Washington, DC, and we see now that Paris is drifting apart from Washington by 30 cm every 10 years. Although no longer in doubt, such a displacement may seem insignificant, but it adds up to 3000 km in 100 million years; continental drift can no longer be disputed.

The break up of Pangaea, which separated Africa from the lower part of North America, began 200 million years ago; it continued with the split of South America from the lower part of Africa 150 million years ago. Greenland split from North America first, then from Europe 80 million years ago.

The first dinosaurs began to diversify and multiply in the immense equatorial rain forests of Pangaea, about 25 million years before Europe began to split from North America and thus formed the Atlantic Ocean. The two major orders of dinosaurs differed from each other in the bones of the pelvic cavity: the '*sauropods*' had a lizard's pelvis, whereas the '*ornithopods*' had a pelvis like a bird's.

The sauropods formed a series of species that were in general large herbivorous quadrupeds, of which some, like diplodocus, reached up to 30 m in length. The order also included some herbivorous bipeds. The ornithopods displayed a large number of biped species, most of them 5 or 6 m long; they also included much smaller species; in adulthood, some are only the size of a chicken. There are also quadruped species such as stegosaurus, with its large bony dorsal plates; because the plates contained blood vessels, we assume they were used to control its temperature – cooling in the shade or warming in the sun.

Several other clues suggest that at least some species of dinosaur began to stabilize their blood temperature by internal processes of regulation, leading to the first warm-blooded animals before the birds. The birds themselves derived from a small species of dinosaur that developed feathers instead of hair. Archaeopterix (Figure 6.4) is an example which appeared about 200 million years ago.

The southern part of Pangaea began to break up into smaller continental plates more than 100 million years ago. In this epoch, Antarctica and Australia split apart. The Australian dinosaurs lived at the time when southern Australia was below the antarctic polar circle, and they had to adapt in order to survive during long polar nights. We assume they survived because they were already warm blooded or, more accurately, homeotherms. In homeothermy, the internal temperature of the body is stabilized by thermoregulation, whatever the outside temperature (within limits). Thus, the speed of biochemical reactions remains

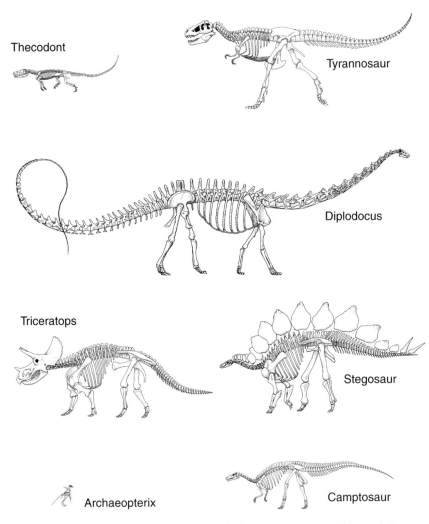

FIGURE 6.4   Sketches of a few dinosaur skeletons reconstructed from their fossils. *Thecodonts* compose a family of several species of reptile, large lizards with very long hind legs; they are the ancestors of all the dinosaurs. *Tyrannosaur* and *Diplodocus* are sauropods. The former is a biped carnivore; the latter is the largest known of the quadruped herbivores, its length being more than 25 m and often as much as 30 m. *Triceratops* and *Stegosaur*, like the *Camptosaur*, are ornithopods; they have the pelvis of the bird, foreshadowing *Archaeopterix*, the first feathered bird.

the same, providing a larger degree of independence from the environment. It implies a greater consumption of food to heat the body, but it also makes more energy available for motion. Not only does the animal run faster, but the extra amount of energy allows it to learn how to fly.

All of this had become possible thanks to the increased amount of oxygen in the air, produced and maintained by the luxuriant vegetation that covered the continents. I have emphasized how much oxygen is an anomaly in chemical disequilibrium with the rest of the biosphere. Plants produce oxygen as a by-product of their food supply; animals breathe it, thus becoming 'parasites' of the vegetation, on which they depend for survival. The 'burning' of their food by oxidation supplies a concentrated source of energy, and the 'dissipative features' (see Chapter 5) of animals demand a larger and larger flow of energy.

Since the flight of birds has been touched on, let us also mention a factor that played a decisive role in the selection of vertebrates that were going to learn how to fly. In contrast to the dense bones of mammals, the bones of dinosaurs were light and porous. These light bones not only allowed them to reach very large sizes but also led to several families of flying dinosaurs (pterodonts), before their scales turned into feathers as in archaeopterix (Figure 6.4), which is considered to be the first 'real' bird fossil.

Two important events occurred during the reign of the giant saurians that could easily have passed unnoticed but instead had far-reaching consequences. The first is the appearance of the first flowering plants and the second the emergence of mammals.

Flowering plants spread through the world 75 million years ago and partially took the place of the large forest of tree-like ferns, conifers, cycads and ginkgoes. Their propagation was due to the spread of pollen carried by insects, which fed on the flowers' nectar. This symbiosis between insects and flowering plants secured the steady spread and survival of both.

The common ancestor of the mammals can be traced back to 250 million years ago. This was a reptile, cousin of the thecodont. Its descendants, 200 million years ago, had become little nosey animals that began to look very much like the little mammalian rodents of the present time. They probably had to protect themselves within burrows, and ventured forth only very cautiously in the silent night, because they were the prey of choice for the dreaded carnivorous dinosaurs. Their reproduction began at first by means of incubation of the embryo in an abdominal pouch, as the marsupials do; but they fast became viviparous.

The breakup of Pangaea and the splitting up of the different continents finally left the offspring of the first marsupials in South America but mainly in Australia, whereas the true viviparous mammals survived throughout the north. They no longer had scales, but hair. Their fertilized eggs developed within the female, whose special organ, the placenta, nourished the embryo during its development.

After birth, the mother continued to feed her young with her milk. These features are valuable for survival in an environment filled with enormous carnivorous animals, from which hiding was the only possible defense.

But a great event was going to end the Cretaceous; 65 million years ago, a cataclysm destroyed a large part of the fauna and the flora, including in particular, all the dinosaurs. However, the mammals survived and the disappearance of the dinosaurs offered them endless possibilities.

What happened?

The major extinctions

The essential feature of natural selection by survival of the fittest implies that the species least adapted to changing conditions are constantly eliminated. Hence there have always been continuous extinctions of species throughout the geological periods. What surprised geologists was that, superimposed upon the steady extinctions, there were times when the extinction rate suddenly reached very high values for a short period, so-called *major extinctions*.

Any event that alters the environment may unloose a chain of consequences that can cause extinctions. Examples are competition with a newly emerged species, change of climate, change in the composition of the atmosphere, large variation in the level of the oceans, exhaustion of a food supply, poisoning by noxious food, intense cold of glacial periods, or intense heat of a runaway greenhouse effect. A series of constantly variable random conditions may occasionally reinforce each other by an accidental coincidence and could trigger a major extinction.

Figure 6.5 has been limited to three types of event that can be major causes of extinctions, namely glaciations, epochs of major volcanism, and impacts of large objects falling from the sky. Figure 6.5 lists only those impacts which are well identified in geological layers large enough to form craters at least 30 km diameter. There are too many known smaller impacts to fit on the figure. Likewise, it is not possible to summarize all the other events that could have modified the environment, like the drift and collisions of continental plates that may have caused earthquakes, or even the changing turbulence of the atmosphere that may have induced hurricanes, and the change in the distribution of wet zones and deserts.

In the geological layers, an event that lasted for fewer than 1 million years can scarcely be distinguished from an instantaneous event like the impact of a comet; and the simultaneous existence of a large number of changing factors makes the interpretation more uncertain. However, a number of major extinctions have been studied in sufficient detail. They occurred during the last 600 million years; that is, during the period for which there are enough fossils.

For instance, the first major extinction took place 440 million years ago, at the end of the Ordovician, when 12% of the number of families identified in the earlier geological layers suddenly disappeared from the later geological layers. The second major extinction dates from 360 million years ago, at the end of the Devonian, with 14% of the families disappearing. Until then, extinctions concerned only marine species, because there were already tens of thousands of species of marine animals and plants, whereas the first land plants without roots, leaves and stems had just emerged in the marshes.

The third major extinction is by far the largest on record; it occurred 250 million years ago, at the end of the Permian. This time, 52% of the shallow-water marine families disappeared, and 49% of the four-footed animals (terrestrial and aquatic) were suddenly eliminated; 8 of the 27 orders of the Permian insects were also annihilated.

The fourth major extinction dates from 210 million years ago; it marks the end of the Triassic. At that time, 12% of the shallow-water families and 28% of the land and marine families of quadrupeds disappeared.

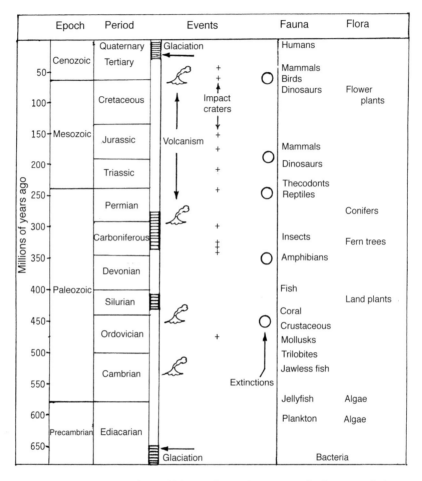

FIGURE 6.5   Recent geological/climatological events in the history of the
Earth. The names of the geological epochs and periods are indicated as
well as the ages given in millions of years ago. Major glaciations are
horizontally hatched; major periods of volcanism are illustrated by
small volcano symbols; impact craters larger than 30 km in diameter are
marked by plus signs; and major extinctions of fauna and flora are
denoted by circles.

Then we come to the last of the great extinctions; the one that caused all dinosaurs to disappear 65 million years ago. It marks the separation of the Cretacous from the Tertiary, which the geologists usually abbreviate by K/T (K comes from the German *Kreide* for chalk, because the C is reserved for the Carboniferous). This extinction saw the sudden disappearance of 14% of the extant quadrupeds (which also included dinosaurs). Families are counted instead of species, because the fossil remains of different species are sometimes too much alike, and the statistics are better with families.

Needless to say the causes of many major extinctions remain controversial, because quantitative data are missing on a large number of confounding factors, i.e. ones that interfere. In contrast, however, the cause of the last major extinction of 65 million years ago has been particularly well elucidated, thanks to the convergence of a number of favorable circumstances.

The American physicist Luis Alvarez, 1968 Nobel laureate in physics for his work on resonance states of elementary particles, and his son Walter, who is a geologist, have set out the whole story. In 1978, Walter Alvarez brought back home a piece of rock the size of a pack of cigarettes that he had identified at Gubbio, in the Appenine mountains of central Italy, as representing the K/T separation of the Cretacean from the Tertiary (see Figure 6.6).

This separation is revealed by a layer of dark clay, 30 to 40 mm thick, that was laid down on the bottom of the sea 65 million years ago, on a layer of whitish limestone. A magnifying glass reveals that 95% of the whitish limestone is made up of little compacted shells of Foraminifera (tiny protozoa living in the sea) mixed with 5% chalk, whereas the upper limstone layer is red and contains practically no shellfish.

Walter Alvarez asked his father how they could compute the time span needed to form and deposit the K/T layer of clay. Luis suggested measuring the content of siderophile metals, because this would reveal the amount of meteorite dust fallen from the sky during the formation of the K/T layer. I have already mentioned that the siderophile metals (some 20 heavy metals such as platinum) were carried down with the iron grains to form the Earth's core. What remains on the terrestrial crust come from recent dust arriving from space. The easiest siderophile metal to measure was iridium, because it readily captures neutrons, thus making the analysis much simpler.

FIGURE 6.6   The K/T separation of the geological layers in the Appenine
mountains in Italy. A coin 25 mm in diameter is shown to indicate scale.
The K/T limit is the layer of dark clay about 40 mm thick that separates
the white Cretacean chalk, filled with Foraminifera, on the lower right,
from the red chalk dating from Tertiary, at upper left. The Foraminifera
are marine shellfish of the Cretacean; they were extinct in the Tertiary,
as witnessed by their absence in the upper layers.

When researchers in the Berkeley laboratory made the analyses, a
totally unexpected and surprising result was found: for an equivalent
volume, the K/T layer contained 300 times more iridium than the adjacent layers! Luis Alvarez concluded that this extraordinary result had
to come from an extraordinary astronomical cause. Since the K/T layer
exists all around the world, Luis Alvarez computed that the mass of the
K/T layer corresponds to a cometary nucleus about 10 km in diameter.
Its collision with the earth must have thrown cometary dust into the
stratosphere. This dust must have remained in suspension in the upper
atmosphere for at least a year or two, making the daytime sky as black
as midnight. This would have stopped photosynthesis for all plants,
and starved animals until they died. Indeed, all animals weighing more
than 20 kg are absent from the later fossil layers.

On the other hand, little animals like the first mammals, which were rodents, were probably able to survive because they could feed on grains and roots for a very long period, and hide from the cold even when most of the plants had withered and died. The presence of this opaque layer of ultrafine dust in the upper atmosphere cut the sunlight enough to lead to a severe winter lasting one to two years, suppressing the cycle of the seasons and causing the average temperature to fall considerably everywhere. Let us imagine that, without any warning, all fauna and flora are taken to Antarctica for more than a year! It is surprising to see that so many plants and animals survived.

Needless to say, in spite of its merits, the Alvarez' theory was not immediately accepted without verification. Some ten years later, about 50 cross-sections of the K/T layer identified in 32 locations all over the world, confirmed in full the abnormal content of iridium and of other siderophile metals. Geologists have tried several other explanations such as gigantic volcanic eruptions, or transit of the Earth through a cloud of cosmic dust; but in May 1984, an unexpected piece of evidence came to light, which confirmed the hypothesis of a collision with a comet.

While studying the K/T layer in Montana, the American B. F. Boher and his collaborators first verified the abnormal content of iridium (it was 200 times higher than in adjacent layers). They then separated the clay from the quartz, by chemical treatment; 10 kg of clay from the K/T layer yielded 1 g of microscopic quartz crystals. Looked at through an electron microscope, each quartz crystal showed numerous fractures filled with silica glass coming from the split crystal. Since split quartz can be obtained only under a pressure greater than 10 000 atmospheres, it must have been the signature of a cometary impact. Such quartz grains fractured by shock and immediately fused again have never been found anywhere except at the site of large-impact craters, like that of the Meteor Crater in Arizona. Thus, an enormous collision took place, which developed a pressure of more than 10 tons/cm^2. There is no possible alternative explanation.

Since then, other studies have indicated that an object, not of 10, but of 30 km in diameter struck the Earth exactly 64.4 million years ago. The location of its impact crater has also been positively identified. It is at Chicxulub near Mérida, in the Yucatán peninsula, in the southeastern part of the Gulf of Mexico (Figure 6.7). The Chicxulub crater reaches 300 km in diameter, and has been dated with precision by

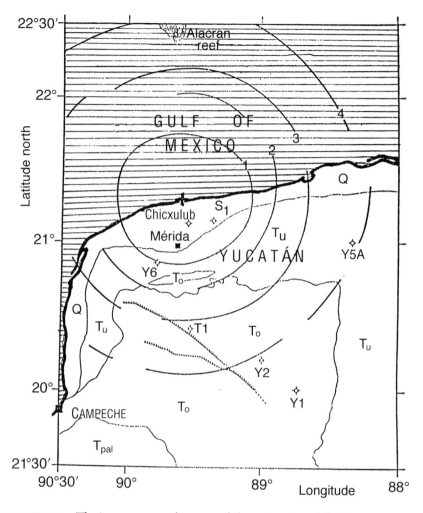

FIGURE 6.7   The impact crater that caused the extinction of the dinosaurs. This 300 km crater is located on the coast of Yucatán and partially under the Gulf of Mexico. It has been dated at 64.4 million years old, and the impact left four concentric circles, detected by gravimetric anomalies and confirmed by soundings at the places identified by the letters. The object that caused the impact had to be at least 30 km in diameter. The crater is deeply buried in layers of sedimentary deposits. Its center is to the north of the city of Mérida.

different techniques. The geologists speak indifferently of an asteroid, a comet, or a meteorite. The probability of a collision with a comet is much larger than that with an asteroid, and a giant meteorite is indistinguishable from an asteroid. Moreover, a comet explains the extinction better, because it is made of a conglomerate of ices and ultramicroscopic dust which would immediately produce a dense fog that would change the climate for a couple of years.

Besides, it was this dust that contained the iridium that settled in the K/T layer. The fallout was slow because the dust was so fine. The particles had enough time to be carried along by the horizontal stratospheric winds, that finally distributed them all over the Earth. The minor quartz crystals were blown away from the impact crater by the gigantic blast.

We have not mentioned the dozen minor extinctions that occurred during the last 250 million years. Only the most recent ones have been identified with narrow geological layers showing abnormal concentrations of iridium. For instance, there is a 39 million year old layer that separates the Eocene from the Oligocene; it contains microtektites originating from North America. Tektites are grains of natural glass, rich in silica and aluminum oxides, coming from a terrestrial sand that has been melted by a cometary impact. The tektite ages show that there were three impacts 38, 37 and 32 million years ago.

Was there a 'comet shower' at regular intervals? Possible mechanisms have been proposed to produce such a periodicity. For instance, could the Sun be a binary star? If it were, the Sun's companion would have to be a star of small mass and faint luminosity, which would now be too far away to be easily identified. Turning around the Sun on a very elongated orbit, it would come back periodically, for instance every 26 million years. During each of its returns it would go through the Oort cloud, from which it would detach thousands of comets, and some of them would bombard the Earth.

This is an ingenious hypothesis that has never been confirmed. I myself took up the task: in 1985 I studied the correlations between all the orbits of the 126 'new' comets known; that is, comets coming from the Oort cloud for the first time. From the asymmetry of the orbits, I computed the trajectory of the putative massive object that perturbed these orbits in order to send them toward the inner Solar System. The results fit in with a small neighboring star, only 30 times the mass of

Jupiter, which passed through the Oort cloud very slowly, about a dozen million years ago. Since this star has never been observed, it must have gone across the Oort cloud only once, and it is not linked gravitationally with our Sun. Moreover, the quasi-periodicities proposed for the major extinctions by some geologists are strongly irregular and are explained just as well by random events. Blind chance sometimes mimics quasi-periodicities that are 'nearly' regular.

A very large number of impact craters are known on the Earth, and many could have been linked to partial or local extinctions, but so far geologists have not been able to establish any conclusive evidence. For instance, 65 impact craters of more than 10 km in diameter have been dated; they were all within the last 600 million years. Only 11 of them were larger than 30 km in diameter; they have been indicated by plus signs on Figure 6.5. Among these craters, the largest known so far is the one recently studied on the Yucatán peninsula, and its age corresponds exactly to the extinction of the dinosaurs.

It was the American geologist Virgil L. Sharpton who recently conducted, with some 10 collaborators, a thorough on-site study of the Chicxulub impact crater. The first results were published in September 1993. The investigation was not simple, because the crater is now buried under a Tertiary layer of carbonate sediments whose depth varies from 300 to 1100 m. They discovered that the impact basin has concentric rings, of which three are complete and one is incomplete and fragmentary; these rings look similar to the multi-ring impact craters observed on other planets.

The largest of these rings has a diameter of 300 km. The rings are remarkably easily detected through gravimetric anomalies, and have been confirmed by borings at different locations. More than half of the crater lies under the Gulf of Mexico. With a diameter of this size, the crater must be the result of one of the biggest impacts to occur not only on the Earth, but also in the inner Solar System over the last 4 billion years – since the end of the intense cometary bombardment illustrated in Figures 5.1 and 5.2. The only known more recent crater of comparable size is the Mead Basin on Venus. It has a diameter of 280 km. It is likely that the Earth has experienced only one impact of this magnitude since multicellular life emerged 1 billion years ago. The object involved was 3 times bigger and 30 times more massive that the Alvarez' early prediction. It is now understood that an impact of this size can explain the features of the K/T layer, as well as the mass extinctions at the end of the Cretaceous.

Other very large extinctions have recently received much attention. In particular, it has been recognized, thanks to the work (published in 1993) of the American paleontologist Douglas Erwin, that the great extinction of 250 million years ago was more significant than had been imagined. More than 80% and perhaps up to 95% of the marine species disappeared at that time; and on the continents, the insects were decimated. It is the only extinction ever to decimate insects over the 390 million years of their existence.

The more this extinction is studied, the more it is confirmed that it was in no way gradual. Paleontologists are beginning to believe that it really was sudden. Since the evidence of a large impact crater dating from that epoch had not been found, other hypotheses began to be considered in earnest. Perhaps a large drop in the sea-level was instrumental? Such a sea-level drop is known because of the large glaciation during the Carboniferous (see Figure 6.5). An even better chronological coincidence was given by the violent volcanic eruptions that produced the gigantic lava fields of Siberia, and built 2 million cubic kilometers of lava. Volcanoes can change the climate when they spew large quantities of dust into the stratosphere. However, another large impact crater has just been discovered, which dates from 250 million years ago. It is 250 km in diameter, and its outer rim is visible in the form of a ring fragment, in the southern tip of South America. Thus, it is not impossible that it played the dominant role in the extinction that marked the end of the Permian and opened the door to the emergence of the dinosaurs.

The study of this large impact crater has not yet been completed, so that it is premature to try to offer a final identification about the phenomenon that produced the great Permian extinction. Let us stop here without drawing any conclusions; rather, let us examine what has become of the small scurrying creatures that survived the extinction of the dinosaurs.

I mean the mammals.

7

THE AWAKENING OF
INTELLIGENCE

Instinct, intelligence and wisdom are inseparable; they are
integrated, react and are blended in hybrid factors.

Alfred North Whitehead, *Adventures of Ideas,* 1943

The fabric of this world is woven of necessity and chance;
Man's reason takes up its position between them and
knows how to control them, treating necessity as the basis
of its existence, contriving to steer and direct chance to its
own ends.

J. W. Goethe, *Wilhelm Meister's Apprenticeship,* 1796

The nervous system

Living organisms learned very early on that the detection of changes
happening in their environment was essential if they were to survive.
Some photosynthetic algae had already learned how to move by
wagging their flagellum, in order to look for places where lighting was
optimal. The first multicellular organisms increased their chances of
survival by detecting changes in their own body, and thus coordinate
the response from their different cells.

Contact between one cell and its neighbors developed in the plants
by means of enzyme exchange; that is, in a purely chemical way. It is a
very slow process which leads to *tropisms* (such as the flower that turns
toward the Sun, or the foliage that seeks light). But tropisms are too
slow for animals; quick reactions for attack or defense are essential. A
faster communication system appeared: some cells specialized in

sending swift electrical impulses from one cell to another, as if along an electric wire. The nervous system began to emerge.

The invertebrates first developed a diffuse nervous system in which fibers are distributed in a network dispersed throughout the body. In the jellyfish, this nervous system is used only for fast coordination of the swimming or eating motions. The flat worm, the first invertebrate to show bilateral symmetry, has an embryonic brain formed by two ganglia which extend in longitudinal nerve strands, forerunners of the spinal cord of the vertebrates.

Among the mollusks, the octopus shows the most sophisticated nervous system and, among the arthropods, insects and spiders reach a comparable (although not identical) degree of sophistication. The nervous system functions at the corresponding evolutionary stage were now concentrated in a brain localized in the head.

A nerve impulse is roughly comparable to a single electric current propagating along a nerve. However, its speed of propagation is much slower than that of an electric current along a metal wire, for several reasons. First, the nerve impulse is carried by ions (much larger than the electrons in an electric wire). Then, the ion pulse has to be regenerated chemically at each *synapse* that separates two consecutive nerve cells (or *neurons*). A very long evolutionary time was needed to improve the speed of the nerve impulse. In the jellyfish, it propagates at 4 cm/second, whereas in mammals it finally reached a speed of more than 100 m/second.

The different sensory organs transmit signals that stimulate the brain, which then reacts by sending instructions to the muscles, along other nerves. Information about the environment, sent by different sensory organs, must of course be analyzed, compared and interpreted by the brain, before being used for survival (i.e. for defense, feeding, protection, etc.). Although the sense of smell played a larger role for the first vertebrates, it is the eye that played the major role as soon as it developed.

Like other chance occurrences in evolution, the present eye emerged from the transformation of an organ initially intended for a different function, namely the photosynthetic cell. It used light as a source of energy in order to transform CO_2 into food. It slowly altered so that it used light as a source of information. By linking with a nerve cell, these photoelectric cells were able to send information about the environment to the other parts of the animal's body. These specialized cells were retained among animals that do not use photosynthesis any more. In the flat worms, a number of these photosensitive cells are located just under the skin where they can detect light, but they cannot yet form an image, nor distinguish directions.

The mollusks enclosed these cells in a hollow. At first, this cavity identified only the general direction from which the light came. Later, the cavity's aperture contracted more and more to form a rudimentary lensless camera. A few photosensitive cells at the bottom of this lensless eye now discerned a few different directions in order to identify the origin of the light source.

The slug protected this cavity with a transparent membrane, which in higher animals became an embryonic lens. However, as soon as there was a lens, there was a complex image that had to be analyzed and interpreted, otherwise it would have been useless.

This is the main reason for the coupling of the eye to the brain and their parallel development. It probably explains the remarkable degree of cephalization in the octopus, whose eye is exceptional among the invertebrates. So, the eye seems to have played an essential role in the development of the brain and intelligence. The basic reason is that the amount of information in a two-dimensional image is so large that it demands not only a faster nerve impulse to transmit it, but also a larger and more evolved brain to filter and interpret the message.

The brain developed slowly in the vertebrates. The first jawless fish, the lamprey, had only a small prominence at the end of the spinal cord. The brain was hardly larger in evolved fish (it weighs 1 g in the trout). It was somewhat larger in reptiles and amphibians, but still small even in dinosaurs. The 25 ton diplodocus had a 50 g brain! Even in large carnivorous saurians, in spite of the need for speed to catch their prey, the brain was not yet very large. The 10 ton tyrannosaur had a brain of 200 g. We do not know the speed of the nerve impulse in the dinosaurs, but some of them had a second cervical center in the pelvic cavity, probably to provide sufficiently fast reflexes in their hind legs, especially when the head was located at the end of a very long neck. The coming of the mammals soon hastened brain development.

The age of the mammals

Mammals came from the therapsids, an order of reptiles that emerged 230 million years ago, during the Triassic. The therapsids were very active little carnivores who improved their hunting technique by running with their hind legs squeezed close to the body, in contrast to the lizards. This gait seems to have given them an advantage in their competition with other carnivores. They were faster in attacking their prey as well as in escaping from the giant saurians.

Many of the qualities of the first mammals seem to have emerged from their search for efficiency. For instance, they developed the double circulation of the blood, with a four-compartment heart that permits a more complete oxygenation of the blood. The thermal isolation provided by fur is a precious complement to endothermy. Warm blood allows sustained activity even when the outside temperature is very cold.

Mammalian evolution led to a surprising variety of sizes and behaviors, from tiny shrews to the huge African elephants, not forgetting whales and mammoths. There are more than 4000 species of mammals, usually classified into 20 differents orders: the rodents are the most numerous. Some orders, e.g those of the elephants and horses, reached a greater diversity in the Tertiary than in the Quaternary.

Since the disappearance of the dinosaurs 65 million years ago, the mammals have rapidly dominated the continents. The order of primates, which includes prosimians, monkeys and humans, emerged about 75 million years ago, in the middle of the Cretaceous. It shows the awakening of intelligence and is of greater interest to us because we are a part of this development.

The dentition of the first primates suggests that they were insectivores, although the identifying signs are a subject of controversy; some could also have been rodents. Fifty million years ago, three families could already be identified that would give birth to the tarsiers, lemurs and monkeys. The tarsier has not changed for 50 million years and it survives in Indonesia. It is a nocturnal insectivore that lives in the trees. It looks somewhat like a tiny monkey, with a 10 to 15 cm long body, accompanied by a very long tail. Its enormous eyes are striking; its digits end in flattened, disk-shaped adhesive pads.

The lemur, still found in Madagascar and in the Comoro Islands, has

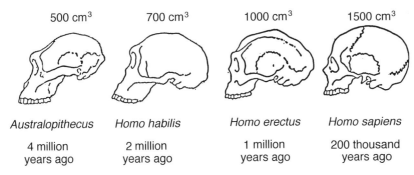

| 500 cm^3 | 700 cm^3 | 1000 cm^3 | 1500 cm^3 |

Australopithecus | Homo habilis | Homo erectus | Homo sapiens

4 million years ago | 2 million years ago | 1 million years ago | 200 thousand years ago

FIGURE 7.1 The evolution of the human brain, simplified, as reconstructed from fossil remains. In terms of geological periods, why has the human brain evolved in such a short time span? See discussion in the main text.

a fox-like face on a monkey's body. It also has a very long tail and enormous eyes, and it measures 15 to 60 cm plus the tail. It eats a little of everything: fruit, buds, leaves, insects, birds' eggs and even small birds.

Among the monkeys, the ancestor of the African proconsul seems to be the common root of all primitive monkeys, for example the little monkeys with long tails of Central America, the macaque and the baboon of Africa, and the tailless apes (chimpanzee, gorilla, orangutan), which are very close to humans.

Australopithecus is a fossil hominid, cousin of the chimpanzee and of the gorilla, who lived between 8 and 2 million years ago and is now extinct. *Australopithecus afarensis*, more than 3 million years old, was discovered in Ethiopia by the American anthropologist Donald Johannson. An almost complete skeleton, reconstructed from its fossil remains, is familiarly called Lucy. Although adult, she was a slender being, of a size comparable to a seven-year old child. The shape of her knee joint implies an upright gait.

Homo habilis appeared less than 2 million years ago, and *Homo erectus*, probably between 1.5 and 1 million years ago. Among these different species, the anatomical differences lie mostly in the size of the skull and in the dentition. Figure 7.1 is limited to comparing, in round numbers, the date of species emergence and the volume of the cranial cavity. In 4 million years, the brain *tripled* in volume; *Australopithecus* had already learned to straighten its gait and to walk erect. Its feet had adapted to running in the savannah and its hands were free to grasp a stick or to cleave a stone. But why had the brain, above all, suddenly developed so quickly?

The evolution of the brain

As I have already remarked, new processes often trigger a branching that abruptly changes the course of evolution. For example, the introduction of sexual reproduction accelerated the rate of evolution because it blended the two parents' chromosomes, exploring all possibilities hidden in the allele forms (the allele chromosomes are similar, but not identical; they form pairs in cells not destined for reproduction).

A somewhat similar story unfolds for the brain. At first, it was used to analyze and make sense of all the signals sent by the sensory organs (in particular, the visualization of images). The brain superimposed on its former functions not only the capability to learn and memorize but also the ability to transmit information to future generations in a cumulative process that we call culture. Culture is the only difference between the nomad hunter and ourselves. It is a sufficiently important phenomenon to justify discussing it in detail.

Transmitting information without screening it did not require a very developed brain. The jellyfish is a good example: its diffuse nervous system serves only to coordinate swimming and eating. The brain is required only when there is too much information, which must then be screened to separate the essential from the incidental. The first problems arose for visual perception. As soon as a two-dimensional image formed in the eye of the octopus, the animal had to analyze thousands of changing dots, many times per second. Competition turned fast in favor of octopuses that developed effective screening mechanisms in a progressively larger and larger brain.

For most animals, vision soon became the tool of choice to analyze and make sense of the outside world; selection helped the survival of the best eyes and of the brains best able to use them. This is why simultaneous development of the eye and the brain occurred in mammals. However, before the eye developed, smell was analyzed in detail by a separate olfactory bulb. This organ is located at the front of the brain in reptiles and frogs, and vestiges of it survive in humans.

Evolution developed the brain in the first primates, probably because most were insectivores. To see and catch small insects trying to escape, not only did their huge eyes need high resolution, but also the brain had to analyze a very detailed and moving image very fast.

Development of the brain in the embryo and fetus of humans retraces rapidly the whole evolution of the vertebrates. Fifteen days after conception, the embryo's brain looks like that of the frog, with an olfactory bulb that protrudes ahead of a very small brain, followed by an enormous optical bulge, and finally by the cerebellum and the spinal cord.

During the following weeks, the core of the diencephalon grows, while the olfactory bulb contracts and nearly disappears. The two hemispheres of the cortex then rapidly become huge and surround the diencephalon. The human diencephalon is the seat of the emotions (pain, pleasure, hunger, thirst, sexual desire) as well as the regulation of blood pressure (which is affected by emotions). It comes from the mid-brain of the reptile and frog, upon which evolution has superimposed successive layers that have developed new functions.

In humans, the optic nerve transmits sensory signals through the thalamus to the cerebral cortex. This is where the images collected by the eyes are analyzed at a level totally inaccessible to the frog, and are translated into a three-dimensional perception where the depth of field is provided by binocular vision. Perhaps it is this analysis of the visual field that gave survival value to the mutations which accidentally enlarged again and again the cerebral cortex of the primates. As seen in Figure 7.1, each step was a jump that enlarged the brain by some 40%. At times, it is reasonable to suppose that there was a sudden overabundance of neurons and dendrites. Not needed for the analysis of visual images, they could be used for a slightly different purpose: the formation of *mental* images which come from recent memories and can help to forecast the immediate future.

Natural selection stimulated the use of short-term memory as a survival tactic. For this reason, it induced the emergence of a feedback mechanism between the memory of events in the immediate past and those of the present moment. This coupling between the immediate past and the fleeting present that is constantly vanishing, produced an unanticipated consequence: that of self-consciousness. It is probable that this self-consciousness emerged at various levels among most primates.

But the large unused portion of the cortex also allowed the development of long-term memory. Its use enabled the memorization of sounds and ended by giving birth to language. Language provides a much more flexible survival strategy than mutations, since language leads to culture, which can adapt much faster to changing conditions. Competition between cultures is in every way the same as competition between species for the survival of the fittest, and it is just as brutal because it translates into wars.

The transmission of culture, at first entirely oral, was deeply modified by the invention of writing, then by printing, which allows the multiple transmission of information without effort or error. Finally, the twentieth century has witnessed an explosion of the diffusion of information, allied with electronic communication, a process whose remote consequences will not be appreciated until late into the twenty-first century.

However, it must be remembered that our whole culture remains and will always remain based on a tradition which, at different times in the near or distant past, ensured the survival of our ancestors. Those who live in present times are all related to their ancestors by an uninterrupted line of tradition that, through time, ensured the survival of the fittest.

The last glaciations

During the last million years, the CO_2 content of the atmosphere was sufficiently low that the greenhouse effect was moderate; glaciations appeared repeatedly as the result of the secular motions of the Earth's axis and the changing shape of its orbit. Glacial and interglacial epochs appeared with a very irregular period, occurring roughly every 60 000 to 100 000 years. Traces of them are visible in deep-sea sediments produced by shells of Foraminifera. Calcium carbonate experienced an increase in the amount of the heavy isotope of oxygen (^{18}O) by about 0.2% during a glacial epoch; it is produced by a well-understood mechanism linked to temperature that we shall not discuss here, but it is used as a dating mechanism.

The latest of these glaciations occurred 600 000, 540 000, 450 000,

350 000, 275 000, 150 000, 70 000, 30 000 and 20 000 years ago. It took an average of 50 000 to 60 000 years to reach a glaciation maximum, whereas it subsided in 10 000 years. The last glaciation 20 000 years ago began long before the previous one had disappeared, so that they blended into a longer glaciation than usual. We are only just emerging from that glaciation, which is not entirely finished, as is obvious by the slow and progressive melting of the Alpine glaciers, of the north polar cap and of the ice cover in Greenland.

The first humans did not suffer glaciations, because they resided in Central East Africa, from Kenya up to Ethiopia. However, they very soon spread to Asia; traces of *Homo erectus* have been found not only in Africa, but also in China and within the island of Java. It was *Homo sapiens* who ventured further north into Europe; fossil traces have been recovered at different European sites (Neanderthal in Germany, Burgos in Spain, etc.).

The evolutionary stages of the species that finally resulted in *Homo sapiens* have been quite well reconstructed: 4 or 5 million years ago, *Australopithecus* left the trees of the equatorial forest and moved into the savannah in search of food; those whose feet adapted to walking erect, survived. About 3 million years ago, the slender *Australopithecus* who survived had the size and weight of an adolescent. He was already using a tool and a weapon: he cleaved a flint stone in two with another stone. The stone edge became sharp enough to slice open the skin of the animals he killed.

Having emerged more than 1 million years ago, *Homo erectus* often saw brush fires that were periodically started by lightning during the thunderstorms which ended the dry season under the tropics. He learned not to be afraid of fire, and somewhere around half a million years ago, he sometimes managed to gather burning fire brands and to keep a wood fire going. He perfected the technique of cleaving flint. Now, he cut it along two faces to improve its slicing edge. The child observed its parents, who maintained the fire and cleaved the flint. He formed a mental image of what must be done; perhaps, a rudimentary language had already appeared to transmit the traditions of cleaving flint and preserving fire.

The adaptation of *Homo erectus* to the nomad life in the African savannah turned out well. In a half-million years, the population increased from 100 000 to 1 million individuals, and competition in hunting forced them to explore farther and farther. The slow change in climate also played a role. The end of the glaciations of 600 000, 540 000 and 450 000 years ago induced a change each time from savannah to desert. We are witnessing such a change in Ethiopia right now, since we are in the end phase of a glaciation.

The point is that nomad groups of *Homo erectus* left in waves for the north, presumably searching for regions with more game. Some also crossed the then almost non-existent Red Sea and went into Asia.

Homo sapiens

Fossil discoveries have not revealed all the details of the branchings that culminated in the evolution of humans. We know, however, that the species *Homo sapiens* had already emerged 200 000 years ago, and that it rapidly demonstrated its superiority in the struggle for survival. Flint cleaving became more refined: 100 000 years ago, fine chips of flint could be attached to a flexible striker, or to a hatchet whose handle was made of bone or wood; 50 000 years ago, other tools appeared (chisels, stone hammers), as well as the first burial places, telltale signs of an emerging culture based on tradition.

Gathering fruits and seeds, and hunting game, there were now 5 million nomads who had spread from Africa into Asia Minor, and then moved out in all directions to India and Indonesia, but particularly through Anatolia to Europe. A handful of them had already reached America by way of Siberia and Alaska. Soon, they invented the use of the bow and arrow. The last glaciations that merged into one another 30 000 and 20 000 years ago, started the extension of the north polar cap down to very low latitudes in Europe. This caused game to become scarce and the struggle for survival became more severe, decimating the first humans who dared to take up abode in Europe.

In temperate zones during the last glaciations, the sky remained overcast at all times, not transmitting enough ultraviolet light to irradiate the lower layers of brown skin. The humans who left Africa had brown skins, colored with melanin, a pigment that protected them from the excess of the ultraviolet light in the equatorial regions. The level of this pigment remained unchanged in the colored races. The result was that, in Europe, the nomad hunters perished in large numbers. With not enough ultraviolet, their ergosterol was not transformed into vitamin D_2 (ergocalciferol), so they suffered from rickets, which prevented normal development of their bones at a time when survival was more difficult than ever because of the intense cold of the winters. The survivors were those for whom an accidental mutation had lightened the skin. This is the origin of the white race in Europe and of the yellow race in China.

About 10 000 years ago, a revolution occurred in the humans' way of life. In the Middle East, some nomad hunters who also gathered seeds, discovered that it was possible to sow in the soil a part of the harvest from the preceding year. They had just discovered agriculture. This implies that they had to settle on the land in order to reap what had been sown. The first villages appeared. The fertile alluvial plains of the Tigris, the Euphrates and the Nile suddenly provided a large excess of food, which gave more leisure to the leaders and also triggered a population explosion. In 5000 years, the population grew from 10 million to 100 million; they invented writing in order to keep records of their crops.

Leisure time leads to inventiveness. The first sedentary farmers discovered new uses for fire: ceramics to make pottery, then the metallurgy of copper, bronze and iron. But the population explosion set off by the reserves of food coming from agriculture continued in an exponential manner, up to the time when saturation phenomena appeared, for crops and fields are limited. At the beginning of the Christian era, there were already 200 million humans. We are going to reach 6 billion at the end of the twentieth century. The twenty-first century will probably see a limit to the expansion of the human race. Further expansion remains uncertain beyond 10 billion.

Another recent revolution has been set off by the explosion of knowledge; the scientific and industrial revolution is less than four centuries old, and it is about 30 years ago that men walked on the Moon for the first time. On the cosmic scale, all of this is so recent that we lack the perspective to appreciate what is happening. The acceleration of progress has been marked by a steady decrease in the number of generations needed to go from one important event to the next (see Table 7.1). One hundred thousand generations were needed for paleolithic humans to learn how to cleave a pebble and to overcome the fear of fire. In this epoch, 10 000 generations are still to come before the discovery of agriculture. Then, 200 generations were required for ceramics and metallurgy, another 200 generations to reach the onset of the

Table 7.1. *The important stages in the evolution of humankind, from the first primates to the twentieth century*
The second column indicates the number of generations from one row to the next. Note how much this number diminishes from one stage to the next. This acceleration comes from the accumulation of knowledge due to culture.

Years of the past	Number of generations	Stages in the emergence of humans
70 million		Among the mammals, the lemurs split from the insectivores; they are the first primates
	—	
65 million		Great extinction of $\frac{3}{4}$ of the species; dinosaurs disappear; mammals survive
	>2 million	
12 million		A primate, the *Australopithecus*, emerges in the equatorial forest of Africa
	240 000	
6 million		*Australopithecus* leaves the equatorial forest for the savannah
	120 000	
3 million		Lucy, a slender *Australopithecus*, measures 1.10 m; her knee has adapted to an erect stance
	40 000	
2 million		*Homo habilis*' brain is larger; he cleaves a flint stone; its edge can cut the skin of game
	40 000	
1 million		*Homo erectus*' brain is larger again; he multiplies; nomadic, he reaches Europe and Asia
	20 000	
500 000		He masters the bush fire; he preserves fire in a fireplace
	16 000	
100 000		*Homo sapiens*' brain is larger again; shaping of flint improves; hatchets and strikers appear
	2000	
50 000		Chisels and hammers appear; the first burial places show the onset of culture and tradition
	1200	
20 000		Still nomads, they gather fruits and seeds; they use bow and arrow to kill game
	400	
10 000		In the Middle East, some discover agriculture; they settle on the land, in the first villages
	200	
5000		They discover new uses for fire: ceramics and metallurgy. Writing, cities and empires appear
	200	
300		Scientific and industrial revolution, with population explosion due to agriculture
	12	
30		Man walks on the Moon; his craft explore other planets and leave the Solar System

scientific and industrial revolution and then, only 10 generations for the conquest of space.

We set out to consider the history of the Universe in its entirety, but the limits of our knowledge have forced us to concentrate more and more on our corner of the Universe. From 100 billion galaxies, our focus has narrowed to only one galaxy, ours. In our Galaxy, we have considered the interstellar chemistry that has developed in one of its spiral arms: the one where we are. We have followed the origin and the formation of our Solar System and, in particular, we have described the formation of the Earth, as well as its evolution during which life emerged. Therefore we have necessarily more and more neglected the rest of the Universe.

Now we must look at what we know less well, and in particular consider the possibilities of life elsewhere in the Universe. The cosmic evolution ranged from complete simplicity to forks and branches leading to more and more features of a growing and quasi-unimaginable complexity. Is it possible that this process also led in other places to forms of life analogous to or different from the many terrestrial forms?

In the Universe we have described, there are at least 100 billion galaxies, each of which contains at least 100 billion stars. Are we alone in the Universe? Even if there are many worlds where life has emerged, are we the only ones to have acquired some degree of intelligence? Are we the first civilization whose technology is so advanced? Or, on the contrary, are we the last to arrive, not yet accepted nor recognized by the 'Federation of Galactic Civilizations'?

These are the really difficult questions, to which we must now try to give an answer.

8

THE OTHER WORLDS

Burchio: Well then, the other worlds are inhabited, just as ours is?
Frascatorio: It is impossible that those countless worlds . . . be deprived of inhabitants similar or even superior
Giordano Bruno, *About the Infinite Universe and the Worlds*, 1584*

The plurality of inhabited worlds

Are there other worlds in the Universe that are inhabited by intelligent beings? This question has always fascinated thinkers and philosophers. In the absence of serious observational data, dreams and wishes nearly always prevail, and most answer yes to the question. The recurrent argument centers by and large on teleology: since the 'reason for the existence of the Earth' is to shelter the human race, the other planets would 'serve no purpose' if they were uninhabited.

In antiquity Lucretius said: 'We have to believe that there are in other regions of space, other beings and other men'. In the sixteenth century, the Italian monk Giordano Bruno 'explained' the plurality of inhabited worlds as God's design and as the purpose of the infinite Universe; for Bruno had read Copernicus and rejected the 'crystalline spheres' of antiquity. He was burned at the stake only ten years before Galileo Galilei discovered the phases of Venus with his new telescope, establishing that planets are not stars, but are 'worlds' like the Earth since they reflect solar light like the Moon does.

When knowledge about the planets became less uncertain, the French man of letters Fontenelle published the famous *Entretiens sur la*

*For his heretical ideas, Bruno was burned at the stake, on 8 February, 1600, on the Campo dei Fiori, in Rome.

Pluralité des Mondes, in 1686. Later, the Dutch astronomer Christiaan Huyghens wrote *Cosmotheoros* on the same subject, published posthumously in 1698. Huyghens believed that even the comets had to be inhabited. This opinion was shared by the French mathematician Jean Henri Lambert, who discussed it in his *System of the World* (1770). The great French astronomer Joseph Jérome Lalande published his *Astronomy for Ladies* where he imitated Fontenelle, then in 1818 undertook to reedit Fontenelle's work, with commentaries. In France, the way was thus open for the romanticism of young Camille Flammarion, who published *La Pluralité des Mondes Habités* in 1862, when he was only 20 years old.

At that time, details of the surface of planet Mars were at the limit of visual observations made with the best available instruments, because atmospheric turbulence limits the perception of fine details, and the photographic plate was not yet sensitive enough. Ever since the observations of Huyghens (1659) and Cassini (1666), the existence of the polar caps on Mars had been known; their whiteness contrasted with the reddish image of the disk of the planet. Mars' rotational period had been estimated to be a little more than one terrestrial day; this was found from the daily displacement of an indistinct triangular spot named Syrtis Major. This spot is the most easily perceived permanent detail on the surface of Mars.

Mars also shows seasonal variations; in particular, the polar cap shrinks during the Martian summer. At that time, the dark spots of the same hemisphere appear to stand out and take on a blue-greenish color, in contrast to the reddish soil of the lighter regions. The seasons are comparable to those on Earth, except that they are longer; the inclination of the axes of rotation is not very different, 25° for Mars and 23.5° for the Earth. At the beginning of the nineteenth century, it seemed logical to interpret the observations as suggesting the existence of vegetation, and therefore of life on Mars. It is thus not surprising that, in 1802, the German mathematician Karl Gauss suggested signalling our presence to the Martian inhabitants, by tracing huge geometric patterns in the snow of the Siberian steppes. Needless to say this proposal was never carried out.

Observations of Mars are much easier when the planet is in opposition; that is, at 180° from the Sun in the sky. At this moment, the Earth and Mars are closer to each other. Because of the rather large

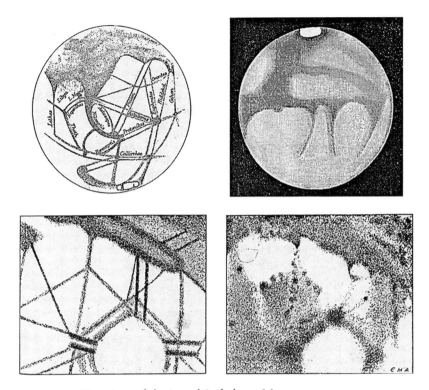

FIGURE 8.1   Drawings of the 'canals' of planet Mars.
Above left: Giovanni Schiaparelli, Milan, June 1888.
Above right: François Quénisset, Juvisy, August 1892.
Below left: details of a small region by Schiaparelli.
Below right: the same region drawn by E. M. Antoniadi.
These drawings suggest that the results depended in part on the draftsman's skill, and in part on the quality of the instrument and the atmospheric conditions of observation. These conditions would sometimes reveal fleeting details that the eye was inclined to blend into the assumed 'canals'.

eccentricity of the Martian orbit, however, there is a time when the two planets are at their closest; this favorable opposition occurs every 15 years and lasts for about six weeks. Starting at a favorable opposition in 1888, the Italian astronomer Giovanni Schiaparelli made a series of drawings of Mars, where he displayed a score of straight lines (see Figure 8.1) that he called '*canali*', the Italian word meaning a channel of water. The straight lines on his drawings were ambiguous, so the translators interpreted them as meaning 'canals' which implies (in French as in English) artificial navigable water ways. Popular imagination was kindled, in Europe as in America, and the press interpreted the observed 'canals' as being the sign of an advanced civilization on Mars.

The American Percival Lowell, a wealthy amateur astronomer and a man of letters, was frustrated that no American astronomer had been able to see the Martian canals described by Schiaparelli. He concluded that most American observatories were poorly located (which was true at the time). He decided to build, at his own expense, a large private observatory at Flagstaff, Arizona. There, the altitude and the semi-desert climate guaranteed the best conditions for observations. On this point he was right, and the large observatories of the twentieth century have followed his example. The Lowell Observatory (as it is still called) was initially dedicated to the assiduous observation of Mars during its oppositions. Lowell, aided by his assistants, often observed there himself. In 1895, he published *Mars* and, in 1906, *Mars and its Canals*; both books were very popular. For 'canals' to be visible at such a great distance, they had to be of an immense width. He explained this visibility by saying that it was due to broad bands of vegetation (like that along the Nile in Egypt). His theory was that the Martians had developed an enormous irrigation system transporting melted water from the polar caps to the equatorial desert regions, because their planet was parched.

In France, interest in the canals of Mars had been kept up so well by Camille Flammarion's work *Planet Mars and its Conditions of Habitability* (1892), that in 1900 the Académie des Sciences instituted the Pierre Guzman prize to be awarded to the person who would find a way to communicate with intelligent beings on a planet *other than planet Mars!* The will and testament settling the Guzman prize deemed that Mars was too easy a target to merit a prize of 100 000 gold-francs! Camille Flammarion himself felt that the exclusion of Mars was a bizarre idea, because it removed from competition the only planet that seemed able to participate.

With observational progress and the improvement in sensitivity of the photographic plates, the existence of the Martian canals became more and more doubtful, since they never appeared on the photographs. The first visual observers were working at the perception limit of the eye; they blended into imaginary lines those fine details that atmospheric turbulence made fleetingly detectable. The best observers who knew how to draw, never portrayed the canals as straight thin lines, but rather as softened shadows. During the first half of the twentieth century, the presumed Martian canals disappeared from the field of astronomers' concern, although they survived in the popular press for a long time.

We had to wait for the Mariner 4 space probe and its first successful flyby of planet Mars, to convince the last incredulous few that the Martian canals do not exist. Instead, it revealed the presence of many craters on Mars, somewhat similar to what we see on the Moon. I felt it was appropriate to relate this curious story in detail, because it shows that observations at the perception limit of the eye are sometimes difficult to interpret and can easily be influenced by desires and emotions.

Exploration of neighboring planets

In the last half-century, the exploration of the Solar System by means of automatic space probes has expanded our knowledge of the other planets and, in particular, of those planets closest to the Earth. It has become clear that observation of planetary surfaces had previously been too much at the limit of previous instruments, and the techniques for studying their physical conditions (temperature, atmospheric composition, etc.) had remained uncertain or non-existent.

At the beginning of the twentieth century, hopes of finding multiple forms of life on many planets in our Solar System were still strong. These hopes progressively faded as our knowledge of the Solar System grew.

Thanks to the detailed study of the planetary surfaces and atmospheres, a new discipline was born: comparative planetology. It slowly became obvious that the emergence of life on Earth had had a stabilizing influence on the climate of our planet. The global amount of vegetation on the Earth influences the CO_2 content of the atmosphere, hence the greenhouse effect, which produces a feedback loop on the amount of vegetation. There are many more positive and negative feedback loops, but the net result is to stabilize the planet's global temperature – as effective a control as that of a domestic central heating system. We now understand the danger of disturbing this steady state, because it depends on conditions that can easily be changed if we are not careful.

This new knowledge has also helped us to understand better the origins of life on Earth, by putting them in their cosmic context. The study of the physical and chemical processes that take place on the other planets has clarified the origin and the evolution of the Solar System, and has explained the processes that have shaped our terrestrial environment, as described in the preceding chapters. In particular, given the conditions of the Solar System formation, it would seem that only rocky planets like the Earth would be able to shelter the emergence of life.

But the rocky planets diverged in their evolution, and we are beginning to understand the basic reasons for their present differences. Let us first consider in detail the cases of Mars and Venus, which are the two nearest rocky planets.

Planet Mars

In 1965, the exploration of planet Mars began in earnest with the NASA space probe Mariner 4, followed in 1967 by Mariner 6 and Mariner 7. As already mentioned, the existence on Mars of numerous impact craters was a revelation. At that time, we were not yet prepared for the discovery, on the solid surfaces of the planets and their satellites, of the scars of the primordial bombardment by relatively small objects such as comets and asteroids. These telltale scars are the 'fossil' traces of the way the Solar System was formed. The only exceptions, where these fossil signs are not easily visible, are in those places where more recent phenomena have hidden them (e.g. by erosion on the Earth, or volcanism on Io).

The first overflights of Mars also established that the large expansion of each Martian polar cap that occurs during the winter of its hemisphere comes from the partial condensation of atmospheric CO_2. Because of the severe cold, this CO_2 forms snowflakes in the atmosphere, which fall down to the ground, making a snow cover which considerably extends the white areas. In the Martian summer, however, the small polar cap that subsists is made of a thick layer of *water* ice.

The Martian atmosphere consists almost exclusively of CO_2, its pressure being about 100 times smaller than that of the terrestrial atmosphere. Because of the condensation of CO_2 into snow, the CO_2 pressure drops in the winter hemisphere, inducing an intense circulation of the gas from one pole to the other, which reverses with the seasons.

In 1971, the NASA spacecraft Mariner 9 was placed in a quasi-circular orbit around Mars (where it will stay for the foreseeable future). Within less than a year, it provided complete photographic coverage of the planet's surface with a resolution near 1 km (Figure 8.2). In particular, this survey disclosed the existence of Olympus Mons, the highest and largest volcano known in the Solar System; it also showed a number of river beds with tributary systems, all dried up, as well as a gigantic canyon. Finally, it revealed that the southern hemisphere consists of mountains and impact craters, whereas the surface of the northern hemisphere is lowland covered by sedimentary layers that hide most of the scars of an early bombardment, suggesting that the northern hemisphere once had an ancient, now completely dried up, ocean.

In 1976, two NASA spacecraft, Viking 1 and Viking 2, were placed in circular orbits around Mars. Each then released a landing module, which landed on Mars at two locations very far from each other, Chryse Planitia and Utopia. The modules photographed the first panoramas of Mars viewed from the surface of the planet.

The principal goal of these landings was to detect whether or not life existed on Mars; not necessarily a form of intelligent life, but any life, even in a very primitive form. The previous chapters have emphasized how the cosmic ratios of the elements very much favor carbon chemistry; this chemistry has selected amino acids and nucleic acids to make life as we know it on Earth. It made sense to choose an experiment designed to detect amino acids, nucleic acids or any organic molecule of a similar type. For this purpose, a mass spectrometer was combined with a gas chromatograph to identify the exact nature of most of the possible organic molecules. Used on Earth, the instruments found amino acids everywhere, even in the middle of the desert or in Antarctica. In contrast, there was no trace of organic molecules on Mars – no amino acids, no nucleic acids of any kind.

Three other series of experiments were intended to detect those functions that seem essential to all forms of life, namely metabolism,

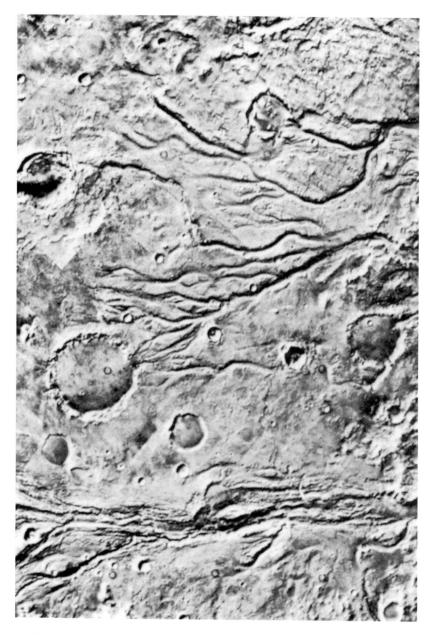

FIGURE 8.2 Ancient dried river beds on planet Mars. These photographs were taken by the NASA Viking space probe, in the Luna Planum area of the northern hemisphere of Mars. The image width is 200 km. The flow

respiration and nutrition. The three experiments gave negative results, despite some ambiguity that was difficult to interpret. In order to feed possible bacteria, the Martian soil had been soaked with a nutritious 'broth' (water plus amino acids brought from Earth). When this was done, gas bubbles came out of the soil. Were there bacteria? No! This was just oxygen from Martian peroxides that frothed on contact with water.

Where did these peroxides come from? The soil of Mars was totally oxidized: in that tenuous atmosphere, the solar ultraviolet penetrated deeply and photodissociated some CO_2, which formed a small amount of oxygen (0.1%) by the reaction CO_2 + ultraviolet = CO + O; this atomic oxygen readily forms peroxides. Incidentally, this oxygen also caused the appearance of an extremely thin layer of ozone in the high atmosphere of Mars; it remains almost undetectable and is insignificant when compared to the layer of terrestrial ozone that protects us from the harmful ultraviolet. On Mars, insignificant as it is, it induces a feedback that stops the formation of more oxygen.

The absence of a larger amount of oxygen in the Martian atmosphere also proves the absence of plants, or algae, or photosynthetic bacteria (as in stromatolites), because their presence would raise the oxygen content very fast. The oxides and peroxides from the Martian soil also explain the reddish color of the deserts on the planet; to put it simply, it is the color of rust. I emphasized earlier that the synthesis of amino acids is incompatible with the presence of oxygen. Let me add that we use peroxides (e.g. hydrogen peroxide) to kill bacteria. The absence of any extant life in the present Martian conditions is thus in full agreement with the environment that exists on Mars at the present time.

This does not exclude the possibility that life tried to start on Mars a long time ago, when the conditions on the early Earth and the early Mars were similar. In the French edition of this book, I had speculated

---

Figure 8.2 (*cont.*)
channels are directed toward the lowest Martian plains, which are covered by sediments and are probably the beds of an ancient ocean. The existence of these channels show that there was once a great deal of liquid on Mars. This proves two points: first, the atmospheric pressure was high enough to keep water in the liquid state; secondly, the temperature was higher than the freezing point of water, proving that the greenhouse effect of a thicker atmosphere was effective.

that a future expedition to Mars might find fossil bacteria deep in dry river beds. The emergence of life should be easy where the proper conditions are met; similar conditions on Mars and on Earth should have brought similar consequences. A comparable bacterial life could have emerged on Mars after the early cometary bombardment, before the later depletion of water and volatile substances. A totally unexpected confirmation of these ideas has recently been announced.

Bacteria in a Martian meteorite

In a press conference on 7 August 1996, NASA announced that fossil bacteria might have been found inside a Martian meteorite. If this extraordinary claim proves to be true, it would be the first evidence that life may have existed elsewhere. It seems to imply that, when the proper conditions are met, life has a strong probability of emerging; this suggests that the Universe should be teeming with life.

Extraordinary claims require extraordinary evidence. First, how do we know that this meteorite really came from Mars? It was found in the ices of Antarctica. During the last decade, Antarctica has become the best place to find meteorites; almost 10 000 meteorites have been found in Antarctica, a factor of 5 more than the number of previous meteorites preserved in museums. The reason for this success is that, when the ice vaporizes, the meteorites remain easily visible on its surface, whereas on other continents, they mix with soil and stones and are not easily distinguishable.

The largest number of Antarctica meteorites came from the asteroid belt, and can be classified as chondrites and achondrites. However, a small number have been identified as coming from the Moon and from Mars. Those from the Moon have been easily recognized by their composition, related to that of the lunar pebbles brought to Earth by astronauts. Only 12 meteorites from Mars have been discovered in Antarctica, and their origin was more difficult to determine. However, the Viking missions carried out detailed chemical and isotopic analysis of the Martian sand and atmosphere. The isotopic analysis of the 12 meteorites showed Martian ratios for the silicates, totally oxidized rocks similar to those on Mars, and even tiny gas bubbles enclosed in the rocks, containing samples of the Martian atmosphere.

The Moon and Mars are the two bodies in our vicinity whose escape velocities are small enough to let rocks escape easily. Hence to find meteorites coming from these two bodies is not impossible. In particular, the collision of a minor body with Mars may throw off ejecta that easily escape Mars' gravity, and are put on many elliptical orbits circling the Sun. A small fraction of these ejecta may remain for a while in orbits crossing that of the Earth. Eventually, a collision with the Earth becomes unavoidable.

Among the 12 Martian meteorites found in Antarctica, only two contained organic matter. The oldest of these meteorites contains small patterns shaped like bacteria. It was shown that it crystallized about 4.4 billion years ago, putting it in the category of the oldest stones that could exist on Mars. It was also established that the impact on Mars that ejected the meteorite in space occurred about 16 million years ago; it finally hit the Earth's atmosphere and landed in Antarctica 14 000 years ago.

In mid-October 1996, it was announced that the second meteorite with organic matter also contained similar patterns. Let us concentrate on the first meteorite, for which more details are known at present.

Four lines of evidence imply that the organic matter present in the oldest meteorite was produced by biological activity of bacteria, when they were alive. Near the supposed bacteria, there are microscopic globules of calcium carbonate ($CaCO_3$), from 1 to 250 μm in diameter; the epoch of the biological activity is given by the age of these globules, estimated to have formed 3.6 billion years ago. They are surrounded by many tiny grains of magnetite (Fe_2O_3) as well as of iron sulfide (FeS). The close coexistence of these grains is a chemical anomaly, consistent only with a biological activity that took place 3.6 billion years ago. They look similar to those produced by anaerobic bacteria in terrestrial fossils; some organic matter in close proximity has been identified as being mostly polyaromatic hydrocarbons (PAH), well-known residues of biological activity.

On the electron microscope pictures (Figure 8.3), these features closely surround the bacteria-like forms. Some are ovoid bodies similar in size and shape to nanobacteria found on Earth in travertine and limestone. Other elongated forms look like earthy filamentous bacteria, although much smaller.

None of these observations is conclusive in itself, but collectively they all seem to support the interpretation that there was a primitive

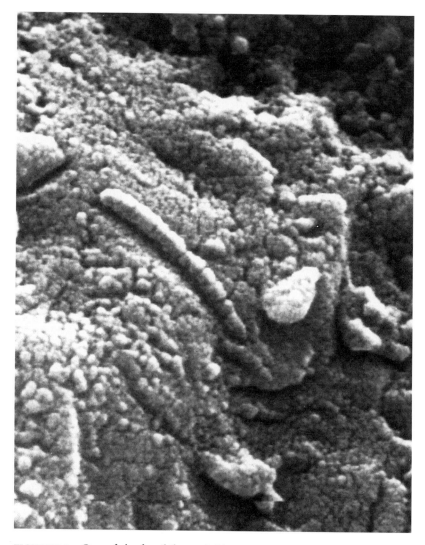

FIGURE 8.3   One of the fossil 'bacteria' in a meteorite coming from Mars. Under high magnification, this elongated form looks like the terrestrial filamentous bacteria found in ancient geological layers of travertine and limestone. The small cylindrical parts look very much like cells, but it has not been possible to verify that there was a cell wall. Indirect evidence stems from by-products of biological activity in the close vicinity of the bacteria.

life on early Mars. These findings require confirmation, in particular from the second meteorite containing organic matter. So far, the 'bacteria' found have been too small and not numerous enough to yield much information. For instance, it was not possible to detect whether any amino acids or nucleic acids were present; neither was it possible to establish whether the 'bacteria' were encapsulated in a cell membrane. Finally, NASA expects to send a sequence of ten mini-missions that will land on Mars during the next decade; the first two were launched in 1996 with spectacular success. These exploratory missions will inevitably take the recent findings into account; in particular, they will try to find fossils in the sediments at the bottom of dry river beds. Return of a sample from Mars is not planned before 2005.

At any rate, preliminary and inconclusive as they are, the present findings bring support to the belief that the emergence of life is not an exceptionally rare phenomenon; on the contrary, when the proper conditions are met, the probability of the emergence of life should be very large.

Planet Venus

Venus is the planet whose size is closest to that of the Earth. Its mass is 81% of that of the Earth; its density is nearly the same (5.3 instead of 5.5), so that its diameter is 95% of that of the Earth. Known as the brilliant Morning or Evening 'Star', it has puzzled humankind since antiquity. Its phases, discovered by Galileo and accessible to the simplest astronomical telescope, are very spectacular. In the early twentieth century, it did not arouse as much interest as Mars, because the impenetrable cover of its changing veil of clouds revealed nothing to telescopic observation, not even its period of rotation. It is only in the second half of the twentieth century, thanks to space probes, that the situation changed abruptly.

In 1958, data from radiotelescopes suggested that the surface of Venus was radiating thermal energy at a temperature near 600 K (about 300 °C). This seemed to contradict common sense. The reflection of radar signals sent to Venus also showed that the planet was rotating in the 'wrong' direction, with an extraordinarily slow period of rotation of 243 days.

In 1962, the NASA Mariner 2 spacecraft flew over Venus for the first time and confirmed that the 600 K radiation was indeed coming from its surface. From 1965 to 1975, the 15 Soviet space probes Venera were launched to Venus. Many missed their target, were lost, or crashed on Venus; at least five landed smoothly on the planet's surface, in spite of difficulties due to the extreme heat. They provided atmospheric profiles of pressure and temperature taken during the parachute descent. The first parachutes had burned up; the more recent ones used an asbestos fiber, and succeeded in safely landing the instruments. The surface temperature was near 460 °C (733 K), even warmer than conditions deduced by radioastronomy.

The atmospheric pressure at the surface was 90 atmospheres; the composition of the air was 96.5% CO_2, 3.4% N_2 and less than 0.1% for the rest of the other gases, including 0.015% SO_2, 0.015% water vapor, and 0.03% O_2. The complete absence of liquid water is normal at this high temperature, but the scarcity of water vapor was a surprise. The clouds at 50 km elevation, which are so opaque that they permanently prevent any view of the ground, consist of fine droplets of concentrated sulfuric acid; their nature was established from their reflection spectrum in the infrared between 3 and 4 μm.

Several American and Soviet spacecraft placed in orbit around the planet used radar to pierce the clouds and see the ground. They drew more and more detailed maps of the planet, revealing a large number of impact craters. The NASA Magellan probe watched two complete rotations of the planet, from 1990 to 1992, to establish (twice) a complete map with unequalled resolution, which varies from 100 to 200 m depending on the region. Besides innumerable impact craters, Magellan revealed the existence of several systems of volcanoes.

How could the evolution of Mars, the Earth and Venus, those three rocky planets that are close neighbors in space, lead to such different temperatures and atmospheres? Since virtually all of the volatile matter seems to have been brought by the same primordial mechanisms, namely a bombardment of comets due to the growth of the giant planets, it is appropriate to discuss the divergent evolution of all the rocky planets together.

Evolutionary divergence of the rocky planets

The rocky planets present in the inner Solar System are Mercury, Venus, the Earth and Mars, to which the Moon can be added because it

Table 8.1. *Escape velocity (in km/second) for rocky planets*

Mercury	Venus	Earth	Moon	Mars
4.2	10.3	11.2	2.4	5.0

is a very large rocky satellite. They are all rocky because they accreted in the same way, in hot zones of the accretion disk, where iron and silicate dust grains had previously lost their volatile fraction. Their present volatile content (atmospheres, water, carbon compounds) is due to the same early bombardment by comets. Why have they evolved in drastically different ways? Let us first consider their atmospheres.

Two factors play a large role in atmospheric retention. The first is the surface gravity on the planet; the second is the temperature at the top of its atmosphere. The gravity of the planet sets the 'escape velocity'; that is, the minimum speed needed to leave the planet and escape into space. This speed is the same for a rocket or for a molecule of gas. The escape velocities for the rocky planets, are given in Table 8.1.

But the speed of a gas molecule is set by the gas temperature; besides, in order to safely escape into space, a molecule must not collide with other molecules, so it has to be already at the top of the atmosphere (the *exosphere*); this explains why the temperature of the exosphere sets its rate of loss.

Finally, at a given temperature, a molecule moves faster when it is lighter. Hence the lightest molecules are lost first; hydrogen, the lightest of all molecules is always lost rapidly. Only the much larger gravity of the giant planets can maintain a hydrogen atmosphere for a long time.

The Moon, Mercury and Mars have an insufficient gravity. But Mars is cooler, hence it retains an (insignificant) atmosphere, whereas the Moon and Mercury were both unable to prevent total loss of their atmospheres over 4.5 billion years. The probable presence of water ice in very small amounts has been recently detected by radar near the pole of Mercury, in valleys perpetually dark and cold. In November 1996 water-ice was probably detected on the Moon, inside polar craters that remain in perpetual darkness. This does not change the fundamental result that no atmosphere or liquid water can survive on the Moon or on Mercury.

The escape velocity of Mars explains why its atmosphere has slowly but steadily escaped into space. Two billion years ago, the Martian atmosphere was still substantial and its greenhouse effect was sufficient to keep Mars' oceans from freezing. Earlier, the atmospheres of Mars and of the Earth must have been very much alike and similar to that of Venus, with an atmospheric pressure, mostly of CO_2, of several tens of atmospheres.

All that remains to be explained is the total absence of water on Venus. The same cometary bombardment that made the terrestrial oceans must have brought an equivalent amount of water to primordial Venus. At that time, however, the surface of Venus was already too hot for this water to condense; it remained as steam that diffused to higher altitudes. Conversely, on Earth, while rising in the atmosphere, water vapor encounters colder and colder regions where it condenses into droplets of liquid water that form clouds.

Our stratosphere provides a 'cold barrier' that stops water vapor from rising any higher. On our planet, the solar ultraviolet that could photodissociate water is stopped by an ozone layer at a height of some 50 km, which water vapor never reaches.

On Venus, the upper-atmosphere temperature was too high to stop the water vapor ascent by the same mechanism. There was no 'cold barrier' to stop steam from reaching an elevation where it is dissociated by the solar ultraviolet. After photodissociation of H_2O, its hydrogen (H_2) was lost from the exosphere, whereas the oxygen molecule was used for different oxidations, including that of the sulfur released in the volcanic gases. SO_2 and SO_3 combined with the last traces of water to make the concentrated sulfuric acid that we now find in the Venus cloud cover.

Has the atmosphere of Venus always been so hot? Since it is nearer the Sun, Venus receives an energy flux that is almost double that received by the Earth. Even 4 billion years ago, when the Sun was 30% less bright, the energy flux reaching Venus was already 140% of that intercepted now by the Earth. We can assume that as soon as the atmosphere of Venus became dense enough, the atmospheric CO_2 did not form solid carbonates because there was no rain; but, in any event, carbonates dissociate on heating to release CO_2. Hence the runaway greenhouse effect we observe now on Venus dates back from the primeval cometary bombardment.

The best proof that this interpretation is correct is the amount of deuterium found by the NASA Pioneer probes in the Venus atmosphere. This heavy isotope of hydrogen is 100 times more abundant than in the terrestrial oceans. Deuterium is twice as heavy as ordinary hydrogen; hence it cannot escape as easily from the exosphere, and therefore it has concentrated steadily on Venus. Its present abundance demonstrates that an enormous amount of water existed early on Venus.

On the Earth, a large fraction of the CO_2 was at first buried in solid carbonates still present in sedimentary layers, because the temperature remained low enough. The emergence of life converted most of the remaining CO_2 into carbohydrates and oxygen; oxygen stabilized the greenhouse effect by a negative feedback. Therefore life regulated the terrestrial 'thermostat', thus preserving the seawater in the liquid state at a moderate temperature. As with all thermostats, regulation is driven by small fluctuations of temperature. These fluctuations come from the movements of the axis and the deformation of the orbit of the Earth, which induce the glacial and interglacial periods.

Elsewhere in the Solar System

In the second half of the twentieth century, biochemists have become more and more convinced that the emergence of life depended primarily on the presence of water *in the liquid state*. At the same time, space exploration showed how difficult it is to find liquid water in the Solar System. Conditions for preserving water in the liquid state are very restrictive, and the Earth is a privileged planet. Under low pressure, liquid water cannot exist; ice sublimates into steam and vice versa without an intermediate liquid phase.

But our ignorance is profound and our hopes and wishes remain strong, so the rest of the Solar System has steadily received much attention. The atmospheres of the giant planets are deep and progressively hotter with depth. The intense gravity of these planets kept a huge excess of hydrogen along with some methane, ammonia and water vapor. At a sufficient depth, steam could condense into droplets of liquid water. Moreover, the Urey–Miller reaction discussed in Chapter 5 could be achieved by those lightning flashes that have been observed on Jupiter; hence many organic molecules could be formed, including amino acids. Cometary impacts could also bring the necessary molecules to the depth where liquid water could exist.

Some scientists have speculated that bacteria could have emerged in the droplets of liquid water held in suspension in the clouds of Jupiter or Saturn. This dream has not been encouraged by the findings of the Galileo spacecraft. After a six-year voyage, Galileo reached Jupiter in 1995, and its atmospheric probe separated from the main spacecraft. It plunged into Jupiter's atmosphere on 7 December 1995 and, for 58 minutes, sent data to the main spacecraft, which sent them back to Earth later. Slowing down its descent by parachute, the probe steadily measured the atmospheric pressure from 0.3 to 23 atmospheres, as well as the corresponding gas density and temperature. The neutral mass spectrometer gave continuous analysis of the atmosphere at different depths.

The first surprise was that in Jupiter's upper atmosphere the helium abundance was much lower than its cosmic abundance. Helium must have been concentrated deeper down, in the same way as the Earth concentrated iron in its core. On Jupiter, it was unexpected because gas turbulence could have been large enough to mix the atmosphere thoroughly.

The important question about possible life has been at least partially answered by the unexpected scarcity of complex organic molecules; the heaviest hydrocarbons found are methane (CH_4) and ethane (C_2H_6). All available nitrogen seems to be in the form of ammonia (NH_3). Water is much scarcer than had been believed. In particular, if present, water should have condensed into liquid droplets in the zone from 6 to 15 atmospheres, changing the temperature gradient. The measured temperature gradient in this zone indicates that there is no trace of water available to condense into droplets.

Of course our only probe may have missed more interesting regions, and Jupiter is huge. Nevertheless, the present findings rather suggest that the giant planets are not favorable places for the emergence of life. A straightforward reason may be that a large amount of hydrogen is not favorable for bringing together *all* the conditions necessary for the emergence of life. The lack of liquid water shatters any hope of making the polycondensations needed for the appearance of early forms of life.

The large satellites of the giant planets may be better places to explore. The four large satellites of Jupiter were discovered in 1610 by Galileo Galilei. For this reason, they are called the Galilean satellites of Jupiter. The largest two Galilean satellites are Ganymede and Callisto: the former is twice as massive as our Moon, the latter 1.5 times as massive. Their global average density is approximately 1.9, which indicates that they are half silicates and half water-ice. They display an icy surface riddled with cracks and impact craters. Without a trace of atmosphere, they have a very cold surface, near -100 °C, and could be completely frozen internally. This at least is the conclusion derived from the early pictures sent by the Voyager 1 spacecraft in 1979. On 27 June 1996, NASA's Galileo spacecraft flew by Ganymede much closer than before. From a distance of only 835 km, high resolving power pictures revealed details down to 70 m. The visible number of impact craters of all sizes is larger than ever. The photographs also show a very large number of grooves that look like faults where tectonic zones have moved away from one another when the icy surface cooled.

But the major discovery was the presence of a magnetic field, implying the existence of a *liquid core*. On Earth, the liquid core is iron, but this is unlikely to be the case on Ganymede, due to its small global density. Any moving liquid with ionized molecules could induce a magnetic field, for instance a deep, salty ocean protected from evaporation by a thick icepack. So, here we might have liquid water again! Only the future will tell whether this interpretation has any chance of being correct. Three more Ganymede flybys have been achieved in 1997, before the spacecraft's attention is turned toward Europa and Callisto.

The two Galilean satellites of Jupiter we have not yet discussed are Europa and Io. They are both closer to the planet and smaller than the previous ones. The nearest to Jupiter, Io, exhibits permanent volcanism. Volcanic ejecta have wiped out every trace of impact craters on its surface; deposits of liquid sulfur coming from its volcanoes constantly reblanket the surface and freeze over quickly. The further satellite, Europa, possesses some recent craters, on a surface of rather smooth ice. In contrast, Io shows no trace of water or ice. The source of its unceasing volcanism lies in the energy released by the tidal forces induced by the constantly changing distance between the satellite and Jupiter, due to the satellite's elongated orbit.

Europa also experiences tidal forces from Jupiter, but they are less strong; not only is its ellipticity not the same, but it is 1.6 times farther away than Io, so the energy from the tides is insufficient to cause volcanism on Europa. However, it has been proposed that this energy would be enough to maintain a fraction of Europa's water in the liquid state beneath a thick icepack. Since 10% of Europa's mass must be water, an icepack a few kilometers thick could protect an ocean about 100 km deep. So this is the second Galilean satellite where oceans of liquid water could exist. Science fiction writers have not failed to fill these hypothetical 'oceans' of Europa with fantastic new forms of life*.

All the bodies too removed from the Sun present the same problem: they cool down severely, because they are not heated enough to compensate for their radiation losses. To prevent water from freezing into ice, there must be a local source of heat. To keep water, even close to its freezing point, from boiling off into steam, there must be an overlying pressure either from an atmosphere or from a thick icepack.

Beyond Jupiter's orbit, no satellite has an atmosphere, with the exception of Titan, the largest satellite of Saturn. Titan is twice as massive as our Moon, and has the same density. It is nearly the size of Ganymede, and its gravity reaches 14% of that of the Earth. It is the only satellite in the whole Solar System that has a thick atmosphere. When the NASA space probe Voyager 1 passed behind Titan, the dimming of its signals was used to estimate the density distribution in the moon's atmosphere, and assess its atmospheric pressure: it is about 1.5 atmosphere. To reach such a pressure in spite of the weak gravity, there must be, per unit of surface area, a gas column 10 times larger than is present in our atmosphere. This atmosphere would extend 10 times farther out than ours.

Voyager 1 also revealed the composition of Titan's atmosphere. The major constituent is nitrogen (90% N_2); the minor is methane (5% CH_4). There are also 0.2% hydrogen as H_2, and 0.1% CO, but only very small traces of CO_2. At the level of some parts per million, there are also ethane (20 ppm C_2H_6), ethylene (0.4 ppm C_2H_4) and hydrocyanic acid (0.2 ppm HCN). At a still lower level, there are traces of diacetylene (C_4H_2), methylacetylene (C_3H_4), cyanoacetylene (NC_3H) and cyanogen (CN).

In a weak gravity like that of Titan, so dense an atmosphere can

---

* In January 1998, the existence of Europa's salty liquid water was finally confirmed by the detection of its induced magnetic field, roughly opposite in direction to that of Jupiter, as predicted by its formation mechanism.

subsist only because it is very cold. Titan's average temperature is only 94 K (-177 °C). Of course no liquid water exists at such an extremely low temperature; even water vapor can barely sublimate from ice, so there is practically no water vapor in the atmosphere.

The photograph of Titan by Voyager 1 is disappointing. The atmosphere is veiled by a dense reddish fog that hides the surface. The solar ultraviolet is probably responsible for this fog, which must come from the polymerizations of hydrocarbons and other organic molecules. It is possible that this fog and organic molecules add up to make a weak greenhouse effect that could raise the surface temperature to -100 °C. Ethane (C_2H_6) has been detected in Titan's atmosphere. Since the melting point of solid ethane is -183 °C, it is possible to imagine oceans of ethane (lighter than water-ice) covering the solid icepack which is assumed to be present on the surface of Titan, but this remains pure speculation.

It is conceivable that Titan's huge atmosphere was derived mostly from the same bombardment of comets as that of the terrestrial planets. The origin of the satellite systems of the giant planets must be small accretion disks surrounding the planets during the building up of their cores; early in their history, the satellites were subjected to the same cometary bombardment that allowed the mass of the major planets' cores to accumulate.

Titan's atmosphere seems to be derived from the chemistry of the comets, and astronomers interested in the origins of life think that a thorough study of Titan's atmosphere is worth while. Not that they hope to find life on Titan, since it is too cold, but because this low-temperature chemistry might give clues to the origin of life that are still missing. This is the reason for the joint effort between NASA and the European Space Agency (ESA) to explore Titan soon. The NASA Cassini space probe is scheduled to release a probe, built by ESA, to penetrate Titan's atmosphere. In spite of delays resulting from budgetary problems, NASA still hopes to launch the Cassini mission at the beginning of the twenty-first century.

Where else could we find life?

As far as we know, life tried to emerge on the two planets of our Solar System where the early conditions were favorable – the Earth and

Mars. It succeeded in stabilizing a biosphere only on the Earth, and aborted soon on Mars because the favorable conditions disappeared early on. Although the exploration of our Solar system is far from finished, our hopes of finding life elsewhere around the Sun have faded away. So where else should we look?

Without solving all the problems, our study of the origins of life have clarified a few prerequisites. First, the nineteenth century hopes that life could use a very different chemistry, even possibly based on different elements, were born from ignorance. We know now that the organic molecules, based on the elements C, N, O and S, are the most abundant available everywhere in interstellar space, and that life has assembled these molecules by the simplest paths, always using the most probable ways and means.

Figure 5.11 displays the three major pathways suggested by organic chemistry to lead to the emergence of life. We have mentioned in particular that the top half of Figure 5.11 already occurs in space, but that the bottom half does not, because polycondensations seem to require liquid water. A condensation reaction is one in which two molecules combine with the consequent elimination of a water molecule. A polycondensation is the equivalent of a polymerization, leading to the assemblage of many small molecules into a long chain, but with the elimination of many water molecules. Biochemists believe that polycondensations require a liquid water phase, because they work by alternate cycles of heating and cooling.

Even if there are other paths, as yet unknown, to make polycondensations, this is a moot point because life has always used the easiest pathways, and we know that the presence of liquid water makes the condensation easier. The large abundance of liquid water on the Earth makes us too easily forget that, although water is the most abundant triatomic molecule in the whole Universe, *liquid* water is very scarce, particularly because it cannot exist in the vacuum of space, since outside pressure is required to maintain the liquid state.

The first conclusion is that life is very unlikely to emerge in space. It can emerge only on a planet where the following conditions are met:

(1) enough pressure to maintain water in the liquid state;
(2) a temperature range between 0 °C (freezing point of water) and 140 °C (destruction of essential organic molecules).

The temperature range mentioned in item 2 may be further restricted by the extant pressure, because liquid water boils away at 7 °C under 0.01 atmosphere pressure, at 46 °C under a 0.1 atmosphere pressure, at 100 °C under a 1 atmosphere pressure, and at 144 °C under 4 atmospheres pressure. The 140 °C limit quoted above remains somewhat uncertain because mechanisms may protect essential molecules to a slightly higher temperature. We can now express the two necessary conditions for a planet to become an abode for life:

(1) a gravity large enough to keep a sufficient atmospheric pressure;
(2) the correct distance range from the central star to maintain the proper temperature.

These conditions neglect the distant possibility evoked for Ganymede and Callisto, namely that liquid water could be maintained for a long time by the pressure from a thick icepack on objects too small to keep a permanent atmosphere by gravity.

A third condition is needed after the emergence of life. In order to develop the full diversity of a biosphere teeming with many life forms, we need time. Judging from the only case we know, we need an immense amount of time, of the order of 4 billion years. This is a serious restriction on the nature of the central star which must keep the planet warm enough, but not too warm, throughout such a time span. Massive stars burn much too fast: a 3 solar mass star leaves the main sequence to become a red giant after only 300 million years! We can now summarize the three necessary conditions to give a planet the most chances of developing a biosphere as we know it:

(1) planet mass, from 0.8 to 4 terrestrial masses;
(2) distance from central star, from 0.8 to 1.6 AU;
(3) mass of central star, from 0.7 to 1.2 solar masses.

In so doing, we have ignored far-fetched possibilities (icepacks over liquid water, giant planets' water clouds, larger greenhouse effects, faster evolution rates, etc.). We have also eliminated the limiting cases, like that of Mars. Even if bacterial life existed on Mars, it was transient and never produced a full biosphere, because the planet's mass was not large enough; it was not able to maintain a large atmosphere and keep liquid water for a long enough time span.

The three previous criteria applied to the Solar System show why the Earth is the only planet with a biosphere. It can be argued that this is a circular argument, because we have used the Earth as our only sample, and we may be too lacking in imagination to think of all the teeming possibilities provided by the rest of the Universe. We must, however, remember how life has always used the simplest pathways and the most probable conditions; statistically speaking, even if there are extant biospheres outside the previous limits, these limits will lead us to explore the most likely places.

Where are the other planetary systems?

Beyond the Solar System, the distances are immense beyond imagination. Only four automatic probes have ever left the Solar System: the two Pioneers (launched in 1972 and 1973) and the two Voyagers (both launched in 1977). It took them about two years to reach Jupiter, and more than 12 years to pass Neptune. At comparable speeds, only three days are needed to go from the Earth to the Moon. These probes have now left the Solar System, and will reach the distance of the nearest stars in about 300 000 years. The length of such a voyage shows to what extent stellar distances defy all our conceptions.

Until now, no telescope has ever been able to show a visual or photographic image of any planet in another planetary system, mostly because a planet does not shine by itself; it only reflects the light of the nearby star. A planet like the Earth is 10 billion times less bright than its parent star; one like Jupiter is 100 million times less bright.

During the last half-century, we have slowly learned that the formation of single stars almost automatically produces the conditions to make planetary systems. This is because, in order to concentrate enough mass, a single star must get rid of an excess of angular momentum. The process scatters matter around in a disk of gas and dust, favorable for planetary accretion (see Chapter 4).

The multiplicity of dust disks detected around young stars is convincing evidence that there must be a multiplicity of planetary systems. Dust disks are more easily detected than planets because, for the same amount of matter, the total surface area able to reflect light is many million times larger, because the mass is spread out in fine dust particles. In order to detect planets, indirect methods must be used.

Table 8.2. *New planets discovered around a pulsar*

PSR 1257+12	Distance (AU)	Period (days)	Min. mass (M)
First planet	0.19	26	0.015
Second planet	0.36	67	3.4
Third planet	0.47	98	2.8

Note:
M is the mass of the Earth.

Recently, the challenge of detecting other planetary systems has been taken more and more seriously by astronomers.

The best indirect method seems to be the detection of *the mass* of the invisible planet. First used for double stars, the periodic deviation from a straight line trajectory of a visible star shows the existence of a gravitational perturbation from an invisible companion. This astrometric method is not applicable to the masses of the smaller planets. A much more sensitive technique is the variation of the radial velocity of the visible star. This variation produces a Doppler shift of the star's spectrum, alternately to the red and to the violet, that can detect the gravitational pull from giant planets like Jupiter. No clear-cut case was identified before 1994; then, the situation changed. The very first case was found unexpectedly by Doppler shifts measured by a different method.

The first planetary system of three Earth-like planets was detected around a distant pulsar. A pulsar, as explained in Chapter 3, is a spinning neutron star, the residue of a supernova explosion, that sends electromagnetic pulses into space. The pulsar PSR 1257 +12 is located in the constellation of Virgo, at about 1300 light-years from us. It spins several hundred times per second and its pulses have been followed over three years with the 300 m diameter radio telescope at Arecibo, Puerto Rico. The pulse detection is so accurate that tiny periodic fluctuations of their frequency have been easily recorded. They were produced by a Doppler effect coming from the alternate approach and recession of the pulsar. These motions were due to the existence of three planets, whose distances to the pulsar, periods around it, and minimal masses are given in Table 8.2. The masses given are minimal values because the inclination of the orbits on the line of sight is not known.

This first detection of a planetary system raises more problems than it solves. First, how have these planets been able to survive the supernova explosion that made the pulsar? Or else, how could these planets have formed after the supernova explosion? At any rate, it is unlikely that any of these planets could be an abode for life, if only because of the dangerous short-wavelength radiations of the pulsar.

Fluctuations in another pulsar (PSR B0329 +54) also suggest the presence of a planet more massive than 2 M at a distance larger than 7 AU. This implies a very long period of revolution (15 to 20 years) and observations over a few decades are needed to confirm its existence.

Since 1994, the patient work of astronomers using the traditional method of measuring the Doppler shifts by spectroscopy has completely changed the situation, with the discovery of several new planetary systems every year. Table 8.3 is a provisional list of a few of the new planetary systems, which all revolve around solar-type stars; these stars remain within the mass limits set by the previous discussion. The precision with which spectral shifts can be measured limits this method to the detection of very large planets.

These detections have become possible because the spectrographs used for radial velocities have been considerably improved. Following the pioneers Geoffrey Marcy and Paul Butler from California, and Michel Mayor and Didier Queloz from Geneva, Switzerland, several teams of observers encouraged by the early successes are now studying hundreds of the nearest solar-type stars. We must expect more discoveries soon.

Of course the next step should be to *see* these planets. This will become possible in the near future. The new infrared camera installed in 1997 on the Hubble Space Telescope offers the possibility of observing giant planets around nearby stars. When coupled electronically, the two Keck telescopes in Hawaii can work as a single interferometer and also achieve the same result. Finally, NASA hopes to put into solar orbit, possibly beyond Mars to escape the glare of the zodiacal light, a set of several telescopes attached together on a long rigid beam, to work as a single interferometer. The two Keck telescopes, used as an 85 m interferometer, will distinguish giant Jupiter-like planets, but a NASA space interferometer should not only distinguish Earth-like objects but study their spectra. Could spectra from the Keck telescopes or from space go one step further, and tell us whether there is extant life on these planets? What is the most characteristic feature of a living biosphere?

Table 8.3. *Planets discovered around nearby stars since 1994*

Star	Planet	Mass (J)	Distance (AU)
51 Pegasi	1	0.47	0.051
47 Ursae Majoris	1	2.4	2.1
70 Virginis	1	6.6	0.45
55 rho 1 Cancri	1	0.7	0.11
	2	5	5
Lalande 21185	1	1.5	10
	2	1.0	2.5
Tau Boötis	1	3.7	0.047
Upsilon Andromeda	1	0.6	0.054

Note:
The planet's mass is given in J, Jupiter's mass; the distance to the central star in AU, astronomical units. Apart from Lalande 21185, the orbital planes are unknown so the masses are minimal.

Oxygen and life

On the Earth, photosynthetic bacteria (see Figure 5.3) appeared very early on and became sufficiently widespread to slowly produce a large amount of oxygen in our atmosphere; 2 billion years later, the process was taken over by algae and plants. There were earlier bacteria that used more localized sources of energy (like sulfur in hot springs); but they never became widespread, and if they still exist, it is in the hot springs of Yellowstone Park, USA, or in volcanic sources deep in the oceans.

Because solar light is the only widespread and permanent source of energy on the Earth, evolution must have favored those bacteria that were able to use it. This must also be true elsewhere; the light of the central star must be the most abundant and permanent source of energy available on other planetary systems.

In the same way, the only widespread and permanent source of carbon available to early life to make food (namely carbohydrates) is the pervasive atmospheric CO_2; but this is already a completely oxidized carbon, and energy is needed to partially reduce it. Fortunately, there is hydrogen in water; but whatever the proportions of CO_2 and

H_2O, there is always too much oxygen to yield the right proportions (H_2:C:O) for carbohydrates. As usual, life has used the easiest pathways and the most probable solutions, by reducing CO_2 to $(H_2CO)_n$ and rejecting the oxygen excess in the air. As soon as this oxygen excess was used up for oxidations in the ground and in the oceans, any new production of oxygen slowly accumulated in the atmosphere.

If the simplest pathway for life is to produce an oxygen excess, then oxygen must be the most characteristic feature of an atmosphere in which life has emerged, because it remains in a permanent state of disequilibrium with its environment.

An automatic consequence of the presence of oxygen is the formation of an ozone layer in the upper atmosphere. Ozone is the triatomic form of molecular oxygen (O_3) produced by the photodissociation of O_2 by the ultraviolet light of the central star. In the near infrared, the absorption band of ozone, between 9.4 and 10.0 μm wavelength, is extremely strong and could become the best criterion for identifying an extant biosphere on a faraway planet.

As an example, the spacecraft Galileo was able to measure the infrared spectra of Venus, the Earth and Mars, as seen from space (Figure 8.4). Such spectra show that extant life is conspicuous on Earth and absent on Venus and Mars. If life disappeared from the Earth, the oxygen, which remains in a permanent state of disequilibrium with the biosphere, would fast disappear through fires and slower oxidations.

Thus there is a real possibility soon not only of detecting faraway planets of the terrestrial type but also of knowing whether life exists on some of these planets. Oxygen seems to be needed for the emergence of animals, which are after all 'parasites' that have learned to live in symbiosis with plants. This was the pathway followed by evolution to bring about the nervous system, the eye, the brain, intelligence and civilizations. So, one should look for oxygen on other planets, but what about intelligence?

Intelligent life

On Earth, intelligence took almost 4 billion years to develop. What happened during this immense period was sketched in Figure 6.2. The first 2 billion years saw the slow entrenchment and evolution of the bacteria and the progressive oxidation of the ground and the oceans.

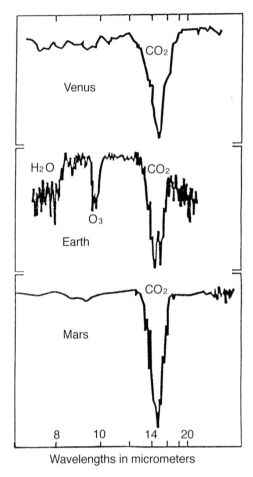

FIGURE 8.4   The infrared spectra of Venus, the Earth and Mars, seen from space by NASA's Galileo spacecraft. The infrared band of ozone (O_3), conspicuous in the Earth's spectrum, is the telltale sign of the existence of life on Earth; it is visible neither on Venus nor on Mars.

The third billion years saw the steady growth of oxygen in the air while the first eukaryotes evolved into marine animals and aquatic plants. The next half-billion years saw plants and animals coming out of the sea and invading the continents; the last half-billion years saw the emergence and the extinction of the dinosaurs, followed by the emergence of mammals and human beings.

The whole process of the emergence of intelligence is based on evolution, and this is an extremely slow process based on random trial and error. In our present state of ignorance, it is difficult to imagine evolutionary shortcuts to intelligence, but this does not demonstrate that intelligence was a highly improbable accident; it just took time to emerge because of the many steps involved.

Was intelligence a mere by-product of evolution without any survival value? This matter was debated by two eminent Americans, the biologist Ernst Mayr and the (late) astronomer Carl Sagan. Mayr believes that intelligence is not an adaptation highly favored by the selection process. Judging from the immense variety of species that occurred before us, Mayr claims that intelligence must be difficult to acquire; he says that it is found in warm-blooded animals only, because brains have very high energy needs. Hence only rare favorable circumstances on Earth have finally produced high intelligence.

Sagan argued that there are so many evolutionary pathways that even if intelligence is unlikely to emerge in each pathway, the total probability is substantial. An overall trend toward intelligence is perceived in the fossil record. As regards the selection pressure, Sagan said that 'it is better to be smart than stupid'. We may have been lucky on the Earth; elsewhere, more than 4 or 5 billion years might have been required to see the emergence of intelligence, but, in the long run, it would have appeared sooner or later.

This book's conclusions rather tip the scales toward Sagan's side. If we look at the long thread going from the primitive bacteria to intelligence, the claim made here is that the bottleneck that slowed down evolution was the need to transform the photosynthetic cell (using light as an energy source) into an eye (using light as an information source).

The photosynthetic cell emerged less than 200 million years after the emergence of life on the Earth. It is understandable that the use of light as a permanent source of energy to produce food had an evolutionary advantage over more localized sources (such as hot sulfurous springs, etc.). Hence it is probable that, on other planets, any bacterial life would soon find a way to use the light of the central star by an analogous pathway.

At the other end of the thread, the crucial step to intelligence occurred recently, during the last 4 million years, when the four species of hominids leading to *Homo sapiens* (Figure 7.1) displayed a 40% to 50% enlargement of the brain, which was repeated three times. Small benefits, e.g. a better analysis of binocular vision, may lead to the selection of a larger brain, whose unused portion will be taken advantage of later, in an entirely new direction. Since the brain changed so fast, its recent evolution cannot be a major bottleneck in the evolution of intelligence.

The major slowing down of the emergence of intelligence came earlier. The problem lay in the extremely large number of very small mutations needed to transform the photosynthetic cell into an eye; the number of steps and the random nature of the evolution make the process very long, even when each step is very probable. The complete pathway includes the transformation of a bacterium into that more complex symbiotic assemblage the eukaryotic cell, presumed to be needed to build the first multicellular animals. The next step was the invention of the nervous system to accelerate the transmission of information among cells. The coupling of photosensitive cells to the nervous system, in order to allow the organism to move toward light, has an important survival value; filtering visual information to understand the meaning of the perceived image was also essential.

Chapter 7 emphasized why the evolution of the eye and that of the brain were coupled, because a visual image contains an immense amount of constantly changing information that needs to be filtered and processed before it can be used. The eye itself has emerged and improved along different pathways, which include the compound eye of insects and the eye of the octopus, rather similar to ours in spite of an entirely different origin, demonstrated by the inverted layers of its retina. The coupling of eye and brain development brought with it a larger brain when the eye improved, as happened in the lemurs. The unused part of the larger brain led quickly to intelligence. We conclude that the slowness of evolution and not its improbability limits the emergence of intelligence; the lifetime of the central star may become the ultimate limiting factor for the emergence of intelligence elsewhere.

In search of other civilizations

The American astronomer Frank Drake proposed 30 years ago to assess the probability of finding other civilizations as a product of several factors that could each be more easily discussed independently. His goal was to evaluate the chances of success of a program of systematic investigation of signals from extraterrestrial civilizations. His motivation arose from new possibilities in radioastronomy. A large radiotelescope, like that at Arecibo, Puerto Rico, was already capable of transmitting detectable signals to the other end of our Galaxy. In a similar fashion, this radiotelescope would also be able to detect signals transmitted by an equivalent instrument located anywhere in our Galaxy. A convenient 'window' of transparent wavelengths of between 18 and 21 cm exists in the interstellar medium.

Frank Drake wanted first to assess the number N of civilizations *extant* in the Galaxy that could have already reached or surpassed our technological level. The time spans involved are immense, and civilizations can appear and disappear. The number N of civilizations that coexist at a given time is given by multiplying the annual rate of emergence of technological civilizations (let us call it r) by their longevity L (the *average* time span between their emergence and their disappearance, for whatever reason). The number of *coexisting* civilizations is then:

$$N = rL$$

For instance, if 10 new technological civilizations emerge *per year* in the Galaxy, and if their average longevity is 1000 years, then there are, on average, a constant 10 000 civilizations capable of exchanging radiowave signals with us.

In order to evaluate the emergence rate r, Frank Drake expressed it by a product of factors:

$$r = SPNBT$$

where S means the formation rate of *stars* per year, P the fraction of stars with *planets*, N the *number* of planets per system, able to shelter life, B the fraction of those planets where a *biosphere* with life exists, and T the fraction of those biospheres that developed a technological civilization. Another factor is often used, although we can neglect it here; it is the fraction D of civilizations willing to talk to us.

The aim of the formula is not to reach an accurate value for r, but rather to assess the role of each factor and judge what it is reasonable to expect. For instance, the stars have formed at a rate $S = 10$ per year on the average, because in our Galaxy, about 10^{11} stars formed in 10^{10} years. The other terms are usually assessed between 1 and 0.1, although some may go down to 0.01 depending on one's degree of optimism. To simplify the discussion, we can accept that each of the factors P, N, B and T is never far from 0.1; hence their product is close to 0.0001 or 10^{-4} and the first formula becomes:

$$N = 10^{-4} L$$

The average longevity of a civilization thus becomes the deciding factor.

If most technical civilizations last less than 10 000 years (from the moment when they can communicate by microwaves) then $N \leq 10^{-4} \times 10^4 = 1$, on average, which means the chances are that we are alone in the Galaxy at the present time. In this time span of 10 000 years, the probability is that no other technological civilization will coexist with ours.

On the other hand, if, on average, a civilization like ours were able to survive for 100 million years, then at this moment 10 000 technological civilizations are coexisting in the Galaxy and are able to communicate with us. Such a survival rate represents of course a very optimistic assessment. Until now, successful animal species have survived approximately 8 to 10 million years.

From the two previous (imprecise) limits, there may be between 1 and 10 000 civilizations coexisting in the Galaxy. In the most optimistic hypothesis, the nearest civilizations would be much more distant than the nearest stars. Statistically, they would be at an average distance of 600 light-years, and we must 'listen' to 10 million stars before having a chance of detecting the first signals. Moreover, after having detected signals, we will scarcely be able to begin a dialogue. No signal can travel faster than light, therefore if we send a signal to a location 600 light-years away, the response will not arrive before the thirty-second century.

In the most optimistic case, practically *all* of these civilizations will be more advanced than ours. Indeed, their *average* life span has been assumed to be 100 million years *after* having reached the technology that we reached 50 years ago. One can fancy that we could learn a great deal from them.

Since exchanges will take a very long time, as soon as we make contact the Arecibo telescope should send them immediately the equivalent of the *Encyclopedia Britannica*, in the hope that they will do the same in 600 years, and we will receive their message in 1200 years. The situation would be somewhat similar to the way our culture has been able to 'communicate' with classical Greece, by rediscovering it and reading books about it at the time of the Renaissance.

Let us now consider the pessimistic case which implies that at this time, we are alone in the Galaxy. There may have been many civilizations, probably thousands of them or much more, but they have all faded away too quickly and practically never coincided in time. The pessimists think that the civilizations that develop their technical know-how learn to manipulate the large natural forces, before controlling their impulses for fight and domination, i.e. residues of their instincts for survival in the struggle for life. In other words, most of them play sorcerer's apprentice before having reached wisdom.

I mentioned how much the emergence of intelligence increased the rate of evolution by changing its process: extremely fast cultural changes, instead of slow random mutations. When different cultures confront each other on the scientific or industrial level, there is no harm if the best prevail, for the benefit of all humankind. But the different cultures too often remain in brutal conflict and are led into bloody wars. It could be argued that, at the beginning, it was just the extension of natural evolution by survival of the fittest. But the means of destruction have gone beyond that limited purpose, and the proliferation of nuclear arms endangers the survival, not only of humankind, but of all forms of life on Earth. If this were the case wherever intelligence has emerged, the pessimists think that there is no chance for technological civilizations like ours to survive over thousands of years. As a consequence, the odds are that, at the present time, we are alone in the Galaxy.

Must we believe that all civilizations come to an end, within a few centuries after they discover radioastronomy and nuclear arms? I do not think so, maybe because I am an optimist by nature. If 10% survive the several critical centuries and finally reach wisdom, this is enough to have many coexisting civilizations in the Galaxy. Who knows? We can still succeed here on Earth; we must recognize that it will not be easy. Perhaps we are at the worst moment; cultures do not evolve at the same rate, and violent conflicts may remain inevitable for a while.

Since Frank Drake proposed his formula some 30 years ago, the search techniques for signal detection from outer space have developed in an exponential way in different countries. The latest techniques automatically explore billions of possible wavelengths and are so fast that more has been done in a day than would have been done in a century by former methods. Until now, Frank Drake and his colleagues have found no indisputable signal coming from an extraterrestrial intelligence. The search has been compared to looking for a needle in a haystack. The comparison is grossly inadequate; the search for extraterrestrial intelligence (SETI) is much more difficult. We should continue to search for at least a century before giving up.

Could a first contact be made by a visit from our or their astronauts rather than by a message? The answer is a definite no; this is a matter of the cost of the energy needed. We can now send a message to the Moon for less than a dollar; a trip by astronauts costs a billion times more! The Moon is quite near, and we must remain aware of the immensity of stellar distances and of the limitations of the energy available to propel spaceships.

Nevertheless, popular belief relates unidentified flying objects (UFOs) to 'flying saucers' of extraterrestrial origin. Why do scientists remain sceptical? Because, so far, no evidence has withstood scientific examination. When frauds or identification errors are eliminated, the 'irreductible' claims are limited to those where information is insufficient and where objectivity is swept aside by emotions and desires. It is reminiscent of the nineteenth century case of the 'canals' of Mars (see Chapter 8). If true extraterrestrials were coming to visit us, the astronomers would be the first to be convinced and the first to be delighted; but the chances are so small that no-one actually expects it to happen.

We have no direct proof of the existence of any extraterrestrial intelligence. Neither have we any direct proof that life exists or existed elsewhere, except perhaps that tentative identification of a fossil bacterial life on Mars. But we are beginning to collect observational evidence that our deduction about the existence of numerous planetary systems around stars was right. From the present rate of discovery of new planetary systems around solar-type stars, the number of planets that could shelter life in the Galaxy is immense. This number must be multiplied again by the huge number of galaxies extant in the Universe.

The most convincing indirect argument lies in the way life seems to have emerged on Earth, by choosing the easiest ways and means. These have always been used by natural selection processes, starting at the most elementary levels, from the most abundant molecules in interstellar space derived from the most abundant atoms made in the stellar cores.

Although the processes leading to the earliest forms of life have not all been clarified, our doubts arise more from lack of knowledge of circumstances rather than from highly improbable conditions. This is the only reason to withhold our final judgment. Otherwise, it seems that the ascent toward a greater and greater complexity that eventually leads to life is an integral part of the general evolution processes of the Universe.

PERSPECTIVES

Man, if he is serious about it, cannot stop from trying to encroach on the region of the unexplorable. In the end, of course, he has to give up and willingly concede his defeat.

Goethe 1832 (to Wackenroder) (quoted by L. Curtis,

Goethe: *Wisdom and Experience*, 1949)

The highest happiness of man as a thinking being, is to have probed what is knowable, and quietly revere what is unknowable.

Goethe, *Maxims and Reflections*, 1832

The evolutionary thread

This book has tried, chronologically, to tell a history of the Universe that began with the Big Bang and continues up to our existence. In spite of many uncertain details and incomplete interpretations, the remaining gaps have not obscured a clear thread of ascent toward a greater and greater complexity, going from atoms to molecules to life, from bacteria to animals to humans, from early cultures to societies to civilizations.

It now remains for us to ponder on the vistas that we have opened up, in order to try to see what they reveal, and to understand the nature of what could still be concealed. Still following the thread of chronology, as long as it remains useful, let us first consider what could have happened *before* the Big Bang.

The 'Augustinian era'

In 1952, George Gamow wittily proposed calling the period that might have occurred before the Big Bang the 'Augustinian era', because Saint Augustine was the first to raise the question of knowing what God did before He created Heaven and Earth. Living at the beginning of the fifth century AD, Saint Augustine was not only one of the first Christian theologians but also one of the most profound thinkers of antiquity. His response to the question was that time is a property of the Universe that God created; hence time did not exist before the Creation.

The flow of time is quite a mysterious property that is neither foreseen nor explained by classical mechanics. The planets could orbit about the Sun in either direction without changing the laws of gravitation. The arrow of time appears in all phenomena that imply heat exchanges, in particular in the form of radiation. The Sun sends us its radiation, but we cannot reverse the flow of time and return its radiating heat back to the Sun.

In Chapter 2, we noted that the matter–antimatter asymmetry led to an asymmetry in the flow of time, as a consequence of the irreversibility in the decay of the K^0 meson. Without that decay, there is no time flow in the world of elementary particles. Before the Big Bang, there was complete symmetry; hence time did not flow.

The well-known British cosmologist Stephen Hawking is one of those who has gone farthest in trying to combine gravity and quanta. In particular, he has shown that, because of the Heisenberg uncertainty principle, the distinction between space and time is blurred when all dimensions of space–time are tiny enough; this was the case at the onset of the Big Bang. This implies that all space–time was then a microscopic four-dimensional hypersphere, finite but unbounded. This means in particular that there is no need to specify boundary conditions: they do not exist. As Hawking says, time then ceases to be defined, just as the direction 'north' is no longer defined at the north pole of the Earth (see Appendix C on this topic).

When there is at last a general theory of quantum gravity, Hawking's intuition will probably be confirmed; this will cut short the mathematical games of some cosmologists who use relativity without quanta. We will have to wait for the twenty-first century to know the answer.

After the Big Bang

The first cosmic process was a quantum fluctuation, which broke the initial symmetry of the tiny bubble of empty space–time and triggered the asymmetry of the four different forces. This set up a change of state which released an enormous amount of energy, launching inflation; later, gravity was felt at large distances, but with considerably reduced intensity.

The quantum fluctuation is a well-known phenomenon. This type of fluctuation triggers the emission of a quantum of light by an atom, or the radioactive decay of a nucleus. It can explain the Big Bang, but the change of state it triggers may be very unhomogeneous. Like water boiling into steam, it may produce 'false vacuum' bubbles surrounded by regions that do not inflate. In this case, many isolated 'baby universes' would have been ejected from ours, leading to a multiplicity of universes (see below).

Without an exhaustive theory of quantum gravity, it is obvious that these ideas are half-baked theories bordering metaphysics.

Branchings toward complexity

The steps toward complexity were achieved by repeated forks and branchings. The first was the separation of the four great forces of nature. The considerable weakening of gravity inflated the microscopic 'Universe' immensely and amplified quantum fluctuations to an extreme, heralding the beginning of large cosmic structures.

Inflation also produced a large amount of matter out of empty space. Each time matter was created, it was surrounded by a field of gravitational potential, whose negative energy transferred mass to matter. Thus, the total energy of the Universe was conserved: it will always remain *zero*. But mass changes in nature. Initially, it was a mass of nearly equal amounts of matter and antimatter, which were easily transformed into radiation, through mutual collisions. A rapid cooling left huge quantities of radiation and a very small amount of matter, whereas all antimatter was wiped out. This brought about the asymmetry that seems to have triggered the flow of time. The residual matter was quenched everywhere into hydrogen and helium nuclei; recombining with electrons, they ended up forming neutral atoms.

Gravity collapsed the cosmic lumps which then formed galaxies. The expansion of the Universe cooled the immense amount of remaining radiation, thus allowing the condensation of the galactic clouds into stars. Gravity compressed the stellar cores enough to reach temperatures of nuclear synthesis. This time, they lasted much longer than during the Big Bang, so that thermonuclear equilibrium was reached and the nucleus of most of the heavier elements could form (carbon, nitrogen, oxygen and 'metals'). With the explosions of the supernovas we reach all the branchings required to make a total of more than 300 different isotopes and about 100 chemically distinct elements, which different processes then redistributed into the interstellar medium.

In the deep interstellar cold, not far from absolute zero, many new branchings appeared, developing all the richness and complexity of organic chemistry. It was triggered by the ionizations caused by ultraviolet light from neighboring stars, and the resulting fast ion–molecular reactions produced a molecular mixture very far from thermochemical equilibrium.

When the later generations of stars formed, their accretion disks now contained silicate dust, metallic iron grains, water-snow and organic stuff of all kinds. When gas turbulence died down in the disk, dust and snows settled into the equatorial plane and ended up forming planets of different types. Silicate and iron grains made rocky planets like the Earth. Further out, snows initiated the onset of larger masses, able to capture gas from the accretion disk before it dissipated.

Still farther from the central star, volatile organic snows, remaining as frost on dust, formed those smaller objects, the comets, a fraction of which ended up bombarding the rocky planets and bringing them the materials needed for the emergence of life. These processes must work everywhere by the same ways and means in the whole Universe, producing the same branchings and the same consequences. However, we only know life exists on Earth; we have no certainty that the necessary conditions are sufficient to trigger the appearance of life elsewhere. This is the reason why the recently found 'evidence' of fossil bacteria in a meteorite from Mars must be carefully confirmed in the future. It is a cornerstone in our worldview that we can neither dismiss nor accept lightly.

We are beginning to have a fair idea of the way prebiotic molecules organize in cycles of non-linear reactions (coupled by catalysis) into 'dissipative systems' able to multiply. Helped by natural selection and the survival of the fittest, these chemical structures selected the best among them fit to survive and duplicate in liquid water. In a way, these 'protobionts' are information-coding machines which duplicate the information that has the best survival value.

These organisms may have found their first source of energy in thioesters; they used the hydrophobic effect of phospholipids to enclose themselves in tiny vesicles, making the first cell wall (for those who need to be convinced that it is easily done, look at mayonnaise through a microscope).

Competition between the first cells soon favored more and more efficient sources of energy, possibly going from the thioesters to the phosphates, then very quickly to the discovery of photosynthesis, since solar light is an abundant source of energy available everywhere on the Earth. It is not surprising that the first known fossils were once photosynthetic bacteria.

The major steps in the history of the Universe are summarized in Table 9.1 to emphasize the major branchings on the road to complexity. Figure 9.1 on the other hand, shows the way in which life evolved on the Earth; the time lapses are suggested in distorted perspective. The most striking feature of this diagram is the evidence for a regular acceleration of evolution, due to the multiple branchings of all the forms of life.

The steady emergence of new forks in the evolution of living beings comes essentially from the thermodynamic disequilibrium of these 'dissipative structures'. The disequilibrium maintains a quasi-stationary state on the margin that separates chaos from perfect order. This is what allows more and more complex structures to emerge. Each living being is a historical accident, depending on all the events that preceded under sometimes necessary, but often accidental, circumstances that go all the way back to the Big Bang. Yet, all these events ultimately produced the emergence of intelligence and led to modern human beings.

In short, this perspective of the emergence of the Universe, then of life, then of intelligence, beginning from nothing – or more exactly, from a small energy fluctuation in the void – seems so improbable, that for many it is totally unbelievable if one does not include a Creator and Designer, who not only set the process in motion, but

Table 9.1. *The branchings toward complexity that led to the emergence of life*

Initial conditions	Cosmic processes	Branchings to complexity
Complete symmetry	Quantum fluctuation	Asymmetry of 4 forces
Inflation	Amplification of fluctuations	Emergence of cosmic lumps
Latent heat	Heating to high temperature	Phase transition to less symmetry
10^{13} K	Quark–antiquark synthesis	Emergence of matter–antimatter
10^{12} K	Quarks condense into p, n and mesons	Phase transition to particles
10^{11} K	Matter–antimatter annihilations	Formation of fossil radiation
10^{10} K	Nuclear syntheses of H, D, He, Li	Ionized plasma of hydrogen and helium
Kinetics at 10^9 K	Very fast cool-down	Abundances are set for light nuclei
4000 K	Space becomes transparent	Emergence of neutral atoms of H, He
Decoupling of matter and radiation	Action of gravity on matter	Emergence of galaxies and their clusters
Expansion of Universe	General cooling	Formation of stars
Stellar cores	Reheating by compression	Syntheses of C, N, O and metals
Spreading of stellar masses	Central temperature varies	Branchings in stellar evolution
Variable fate of stars	Ejection to space of C, N, O and metals	Variable enrichment of galactic clouds
Galactic clouds	Ion–molecular chemical reactions	Emergence of organic molecules
Late generations of stars	Settling out of dust in disk around star	Formation of rocky planets
Cold zones of accretion disks	Clumping of icy grains	Formation of comets
Diffusion of cometary orbits	Bombardment of the planets	Emergence of possible biospheres
Liquid water	Oceans on some planets	Emergence of life

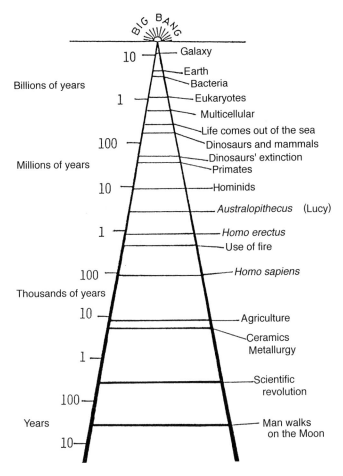

FIGURE 9.1   Emergence of life in (distorted) perspective with respect to the
Big Bang. The perspective is distorted because a logarithmic scale was
needed to include all the durations. A linear scale for the time elapsed
would have boosted the first eons immeasurably, and the recent events
would all have been compressed within an insignificant line. The
logarithmic scale separates all important branchings, and establishes
that the ascent to greater and greater complexity constantly accelerates,
at an approximately exponential rate.

arranged the laws of nature so that the needed coincidences were not chance events.

Since, by definition, metaphysical causes cannot be verified by experiment or observation, some scientists wonder whether it would be possible, in observing the Universe, to identify the coincidences that have occurred to our benefit, in order to evaluate the probability of a chance occurrence. In particular, is the way in which life has emerged on Earth a highly improbable accident, which has never been repeated elsewhere? In the section 'Chance or necessity?' in Chapter 5, I concluded that life was a *very probable* event, because it was produced, not in one single operation from scratch, but in innumerable small steps, each of which was extremely probable.

The question is now extended to the rest of the Universe: given the laws of physics, is life a necessary or a chance event?

The anthropic principle

A score of years ago, the British physicist Brandon Carter introduced the term 'the anthropic principle', in order to express the idea that everything that ever existed before us in the Universe must be compatible with our existence on Earth.

The idea was already in use before the term was coined, because in its simplest form it is so obvious that it is a tautology. The story of the identification of the nuclear reactions that make carbon in the stellar cores is an excellent example of the principle. Table 3.1 shows that the nuclear combustion of helium (^{4}He) into carbon (^{12}C) does not combine *three* atoms of helium instantaneously; this is because a triple collision has a very low probability of occurring. There must be an intermediate stage involving ^{8}Be, as shown in Table 3.1, but the beryllium then formed must not rebound or break down before reacting with the third helium nucleus still required to reach carbon. At the temperature of the stellar core, in order to form carbon the beryllium must be 'in resonance' with helium when they collide. Resonance implies the existence of three close energy levels in helium, beryllium and carbon, to allow the two consecutive 'jumps' from helium to carbon, without stopping at beryllium.

The British astrophysicist Fred Hoyle (mentioned earlier in this book in another context) was working on this question more than 40 years ago. He checked that in the tables of energy levels of the atomic nuclei there was no energy level for carbon at the beryllium–helium resonance. He computed the exact value (7.65 MeV) of the level that should exist in order to make carbon in the stellar cores. Knowing that there was no way to make carbon elsewhere in the Universe, and that carbon exists on Earth, the level he had computed had to be there. Upon his insistence, the Caltech Laboratory of Nuclear Physics in Pasadena, California, agreed to verify his hypothesis and found the resonance at the predicted level. It remains a beautiful example of the application of the anthropic principle.

The *weak* and the *strong* anthropic principles must be distinguished. The weak anthropic principle says that, at the location where we are (on Earth) and at the present time (some 15 billion years after the Big Bang), the necessary conditions for our existence must be met.

The beryllium–carbon resonance implies more: this is the strong anthropic principle, which assumes that everywhere and at any moment in the Universe, the physical laws are set in a way that allows for the emergence of conscious beings such as ourselves.

The *weak* anthropic principle is self-evident but not very useful. It may be used to 'explain' why the age of the Universe is so great. In order for us to exist, the chemical elements present in our bodies had to be made in the cores of stars, then ejected into space, then brought down to Earth. By adding up all the necessary time spans, we are not too far from the estimated age of the Universe. This 'shows' for instance that 6000 years (to fit in with the literal interpretation of the Genesis story) is too short a time.

The *strong* anthropic principle allows slightly different interpretations that remain, however, always at the margin of metaphysics. In its strongest form, it opens the door to the mystery of our existence. If our existence implies the existence of a number of *very unlikely* coincidences in the natural laws, have these coincidences been selected by design, with the purpose of bringing about the existence of humans on Earth? This is a teleological argument in all its splendor, which explains the Universe in terms of a final purpose: the emergence of humankind.

Some go even further: if we manage to compute the probability of each chance coincidence, and conclude that their simultaneous appearance is unbelievably unlikely, this becomes proof that the

Universe has a grand design, oriented towards a definite purpose. Of course, it would be only a probabilistic proof.

We have not yet reached that stage. Anything looking like a coincidence may have a straightforward explanation that will become evident when we know more, in particular when we master a unified theory of quantum gravity. At any rate, we cannot compute with any accuracy the probability of most of the observed coincidences. Let us look at some examples.

First, there is the numerical value for each of the four force constants that were reached when the four fundamental forces were separated during the Big Bang. A slightly different value for the constant of the strong nuclear force (which acts only inside the atomic nuclei) would suppress the beryllium–carbon resonance, and thus the possibility of making carbon in the stellar cores. A slight change in the weak nuclear force would modify the radioactive decay of the neutron, and thus upset completely the abundances of all the elements. A slight change in the electromagnetic force would considerably change the size of the atoms, as well as their electronic structure, completely upsetting molecular chemistry. A change in the force of gravitation would change the sizes and lifetimes of all the stars.

However, we have no proof that all these changes are possible. To us, the force constants seem arbitrarily chosen, probably because we have not yet succeeded in unifying gravity and quanta into one single theory. Such a synthesis may reveal that we have no choice. Even if we assume for a moment that the force constants are and will remain arbitrary, we have not yet proven that a form of intelligent life would not be possible under conditions very different from ours. Life on Earth has always used the must probable ways and the most likely means. Carbon, nitrogen and oxygen were selected automatically because they led to the most abundant prebiotic organic molecules; water was selected because it was the most abundant triatomic molecule in the whole Universe; and so on. If the nuclear reaction of helium had stopped with beryllium, then beryllium would have become very abundant; but what would happen to the possible beryllium chemistry, if the electromagnetic force constant were also changed?

The same reasoning applies if the Universe were of a different scale or density, so that it would be really difficult to evaluate the probabilities of life in all possible universes with all possible values for the force constants. The probabilities of the emergence of different forms of life might be much greater than we can imagine, because of the immense multiplicity of pathways opened up by the changing values of the force constants.

We can conclude that, up to now, the strong anthropic principle does not yet allow scientific arguments to fit in with metaphysical considerations. Its use to establish that there is an ultimate reason for the existence of the Universe is premature. The existence of a grand design cannot be ruled out but, since we have not been able to prove it, the problem is beyond the capabilities of present-day science.

Perhaps the situation will be somewhat clarified when we unify the theories of gravitation and quanta. At that time, we might interpret the meaning of the values for the constants of the four forces of nature, and reduce the number of arbitrary constants; we could then assess the meaning of the coincidences found in the Universe.

Multiple universes

Another possible interpretation of the astonishing coincidences (if in the future, they are proved to be astonishing) would be to imagine that there are actually an extraordinarily large number of universes 'parallel' to ours, and wholly unknowable to us. We mentioned already that the ejection of innumerable 'baby universes' separated from ours is not an impossible consequence of a turbulent inflation. In our present state of ignorance, we cannot rule out that any possible values of the fundamental force constants could exist in these different universes, and life would appear only in those which display the necessary coincidences; the anthropic principle implies that we would be inevitably located in one of these well-chosen universes. The multiple-universe hypothesis removes the necessity of an ultimate purpose, since life would emerge by accident in a universe where the needed coincidences have occurred by chance.

Because these other universes cannot be observed, the hypothesis seems to belong to the realm of science fiction; unverifiable, it seems to have no connection with physical reality. However, this is not completely the case, because the origin of these universes could be explained by a turbulent inflation, and their existence is supported by a possible interpretation of an unresolved paradox of the quantum theory.

Present-day quantum theory rests on solid experimental bases. The equations of quantum mechanics predict with unbelievable accuracy the exact wavelengths of thousands of spectral lines emitted or absorbed by atoms and molecules, as well as most of the phenomena observed in the subatomic world. These equations have never been found to be in error. However, they describe a strange subatomic world that we do not know how to interpret below the level of the observed phenomena.

Three-quarters of a century ago, the founders of quantum theory – the Danish physicist Niels Bohr, the German Werner Heisenberg, the Austrian Erwin Schrödinger and others – held long meetings at the Bohr Institute in Copenhagen, in order to try to understand the meaning of these equations that worked so well. They concluded that it was not necessary to try to understand, for example, why light behaved sometimes like a particle and sometimes like a wave. For them, there was *no profound hidden reality* in the mathematical formulas. This is what is now called the *Copenhagen interpretation*.

Many physicists, among whom Einstein is the most famous, have never accepted this interpretation. The debate continued for 70 years and produced a long list of new interpretations, of which most are not more convincing than that of Copenhagen. Since I will skip over the question here, this discussion must necessarily be only superficial. In 1935, Einstein and two collaborators, Boris Podolsky and Nathan Rosen, proposed that an experiment be tried that would prove that the quantum theory, although exact, is still *incomplete*.

Easy to conceive (Figure 9.2), the experiment was difficult to achieve. Tried first in 1972 in California, then in a more irrefutable form in 1982 by the French physicist Alain Aspect in Paris, it demonstrated that the quantum theory is *complete* and that Einstein was wrong, but it left the door open to two different interpretations of the quanta, each as paradoxical as the other.

The first is that, once two quantum particles have interacted with each other, they can influence each other forever; more specifically, they can 'communicate' faster than light, however widely they become separated. When this principle is used for the Big Bang, it means that the whole Universe remains interconnected; this yields a holistic view

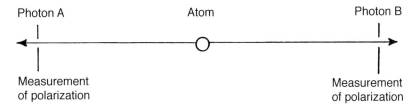

Photon A   Atom   Photon B

Measurement of polarization   Measurement of polarization

FIGURE 9.2   The paradoxical experiment proposed in 1935 by Einstein, Podolsky and Rosen, to test whether the quantum theory was 'complete'. The proposed experiment has different minor variations; here, only one is being described. Two photons are emitted simultaneously by a single atom. When they are far apart, each of their polarizations is measured separately. Polarization is a vector property (this means it has size *and* direction). A vector can always be decomposed into two perpendicular components. The direction of the polarization is not designed until the measurement plane is chosen. Einstein hoped that the two photons would not influence each other's polarization after they were separated, demonstrating that something was missing in the quantum theory. In 1982, Alain Aspect proved that Einstein was wrong. The experiment was made on a large number of photons, but the measurements showed the probability distribution predicted by the quantum theory, which is very different from that predicted by common sense. The two photons seem to influence each other faster than the speed of light. One interpretation claims that measuring A sends a reflected wave which *travels back in time* to B, where it is cancelled by its interaction. Since this wave does not carry energy, it does not contradict relativity (relativity requires that energy cannot be transported faster than light).

of physical reality. This non-locality of individual particles, which is fundamental for all quantum entities, remains totally mysterious. However, it was proven to be true by Alain Aspect's experiment, unless we accept the other interpretation.

The second interpretation is that there are multiple universes. This astonishing claim put forward to clarify the equations of the quantum theory was proposed in 1957 by the American Hugh Everett in his doctoral thesis, supervised by the well-known American physicist John A. Wheeler, then at Princeton. Published in the *Reviews of Modern Physics*, it has been recognized, after careful examination, as a possible interpretation which does not contain any internal error in logic.

Everett showed that the quantum equations can be interpreted by saying that each time a quantum particle is interfered with it is being split. One of the two particles remains observable in our Universe, whereas the other disappears and emerges in a parallel universe. This splitting occurred from the first quantum fluctuation in the Big Bang and innumerable parallel universes have therefore been created.

In our present state of ignorance regarding a general theory that unifies gravitation and quantum mechanics, it can be imagined that these parallel universes have different force constants, and only the universes with the correct coincidences allow the emergence of life. This interpretation removes all purpose to our Universe and thus all metaphysical interpretation about the existence of a Creator.

If Everett's interpretation is accepted it is still improbable that the ultimate theory of quantum gravity would allow the values of the force constants to fluctuate as if it were an effect of each splitting. For my part, I cannot readily accept the presence of parallel universes in quasi-infinite number, merely to interpret the equations of quantum theory. I prefer to accept the Copenhagen interpretation that there is no profound reality, even if the logic of the hidden phenomena escapes us.

However, Alain Aspect's results remain to be explained: for instance, the fact that two photons, separating at the speed of light in two opposite directions, continue to influence each other's polarization as if they were communicating at a speed greater than that of light. I believe that the problem comes from our difficulty in understanding the phenomenon of the flow of time. On the scale of subatomic particles, time does not flow in the usual way, because some phenomena seem to go backwards, against the flow of time. This creates ambiguities, such as that introduced by a negative electron travelling back against the flow of time and thus appearing to be a positive electron.

Another ambiguity of the same type is suggested by Maxwell's equations which describe the propagation of light. They have two formal solutions: a regular wave, and a wave which travels backward in time. The traditional wisdom has been to neglect the second solution. It was said that it had no significance because we 'well know' that no light ray can possibly travel backward from our eye to its source.

However, John Wheeler has always been interested in the interpretation of the strange world of the quanta. Much earlier than his work with Hugh Everett, more than half a century ago, he collaborated with another young physicist who was to become famous: Richard Feynman, who received the Nobel prize for physics in 1965. Together, they discovered that the mystery of the 'collapse' of the probability waves occurring when the position of the quantum particle is being measured could be due to wave functions that travel in both directions, toward the past and toward the future.

In this hypothesis, the receiver of a light signal always produces a 'reflected' wave travelling back in time, along the path of the light ray, reaching the emitter at the moment of the emission; beyond, it interferes with the part of the signal from the emitter which is also travelling back in time. Since the two signals have opposite phases, they cancel each other, removing any trace of the signal *before* the emission. All we see is the effect of the wave travelling with the flow of time.

This interpretation did not seem very useful except to philosophers, because it did not seem to leave any observable effect. This changed in 1982, because it could be used to explain the mysterious results of Alain Aspect's experiment. The two photons, separating at the speed of light, keep in touch and can coordinate their polarization through the phase of the wave travelling backwards in time. This does not transmit energy faster than light, so that it does not contradict Einstein's theory.

For our discussion, another important consequence is that this interpretation does not require other 'parallel' universes. Hence we cannot use their immense number to claim that we are in one of the universes where life has emerged by chance, because of random coincidences in the force constants.

As a consequence, can we now claim that our Universe has been created by design to make life appear? Not yet, because in our ignorance of quantum gravity or any grand unified theory, we have to wait for a future century to understand whether the present values of the force constants result from an internal logic that is still beyond our present-day understanding.

The scale of the Universe

Until now, we have put on one side questions where our uncertainty is very great, in particular the size of the Universe, including whether it is finite or infinite. What we have called *the Universe* or *our Universe* has always been the known Universe, bounded by what we can perceive out to the limits fixed by the cosmic afterglow of the fossil radiation. The accessible Universe has a radius of the order of 12 billion light-years (see Appendix D), and was caused by the inflation following the Big Bang. Personally, I have very few doubts about the existence of inflation, but I admit that this phenomenon is neither well understood yet nor quantitatively well defined, in the absence of what I have called the 'final grand theory' which, in particular, will unify gravitation and quanta.

It is conceivable that inflation could have swelled the primordial Universe up to a size millions of times larger than that which we can perceive. This is in agreement with the observations that suggest that the Universe is not far from the 'critical' density (see the Glossary), and thus very 'flat' (its spatial distortion is very weak). All that we can compute is that part of the Universe which we can observe. The portion which is now observable was 0.002 mm in diameter at the so-called Plank time; that is, 10^{-42} seconds after the Big Bang.

The future of the Universe

We still do not know whether the Universe will have an end in time, or will be eternal. Because of its nearness to the critical density, it has not been possible so far to determine, by observation, how space–time is curved; the mathematical models leave all possibilities open. All that can be said is that the rate of expansion of the Universe is going to continue for a long time, in terms of hundreds of billion years. The nearness of its density to the critical density implies that the expansion could slow down gradually without ever stopping completely.

After the death of all stars, including the least massive ones which will burn the longest, matter and radiation will go down toward an average temperature closer and closer to absolute zero. Because of the expansion, the temperature will continue to fall, so that the 'thermal death' of the Universe due to a perfect thermodynamic equilibrium will never happen. We are already far advanced in this evolution, since the average temperature of the present Universe is close to 3 K. The present stars are only the insignificant embers, not yet completely extinguished, of the original fireworks.

The future of our civilization

Homo sapiens is a species that emerged 200 000 years ago; it supplanted *Homo erectus*, which disappeared for reasons we do not know. Survival of most species rarely exceeds 8 or 10 million years. The usual limiting factors are accidental mutations combined with the more or less catastrophic changes that happen in the environment, as well as the competition between species that leads to the struggle for survival.

By and large, we have barely reached a small percentage of the normal survival lifetime for our species; but the conditions of evolution have changed considerably. Our culture, or rather our cultures, have led to a new type of evolution which, although faster, is not fundamentally different from the one controlled by the struggle for life among animals. In particular, we have invented war between cultures, which is nothing less than a competition for the survival of the fittest, and we use more and more ferocious technical means, of which the proliferation of nuclear arms is only a recent example.

Moreover, our intelligence led us quickly to the invention of agriculture, which ultimately induced a population explosion that has not yet ended. The accumulation of our knowledge, at first by oral tradition, next by writing, then by printing, and finally by electronic means, led to the scientific and industrial revolutions, which rapidly made us the uncontested masters of the biosphere. This conquest of the Earth has led us, almost without our perceiving it, to destroy the environment and, as a result, even the very existence of hundreds of thousands of animal and plant species. We are thus responsible for the largest extinction of living species known on Earth since the extinction of the dinosaurs, 65 million years ago.

A quarter of the population of the globe has thus reached a level of extensive development, and takes advantage of the comfort brought by an industrial support base that exploits natural wealth, while producing a food surplus. Three-quarters of the world's population, on the other hand, lives at different levels of underdevelopment, with numerous regions of malnutrition and famine, and with different cultures that remain constantly in competition.

Even in the most developed countries, this competition between cultures remains harsh and pitiless. It is still a matter of survival of the fittest culture, in the strict tradition of the evolution of species, as witnessed by the elimination of the soviet autocracy, after 70 years of competition with the democracies. By and large, the most ancient cultures are supported by traditions that have secured their survival for centuries. This is because, at the cultural level, traditions now play the role formerly held by genes.

In the same way that the most violent competition breaks out between two neighboring species that have sprung from a common stock, so too with the cultures. In particular, cultures that have split from a common tradition (such as an ancient religion) remain in open competition that often becomes bloody, as between Judaism and Islam in the Middle East, or Hindus and Sikhs in India.

The rapid progress due to the blossoming of the postindustrial era has led to similarly rapid changes in the ways of life and in the system of values among the different cultures of Western civilization. A growing margin of the population seems to have lost the ancestral constraints of tradition that were taught from a very young age within the family, because the traditional family itself is disappearing. We are in a new environment that threatens to upset the traditional bases of Western culture. Is our civilization reaching a peak? Do we go toward its decline? Or conversely are we moving on toward major progresses that will lead to a new world view brought about by the fusion of several cultures? The French essayist Paul Valéry wrote: 'we civilizations, we know now that we are mortal'.

The next millennium

The British philosopher and mathematician Alfred North Whitehead emphasized in 1933 that it is the business of the future to be dangerous. The major steps of civilization are processes that nearly destroy the societies wherein they arise. This is understandable, since it is still natural selection that brutally eliminates the cultures less fit to survive. Cultural changes may occur very rapidly; because a culture is passed on by educating the very young child, the social structures that define a culture can be upset in one generation, or two at most.

Contrary to what the French philosopher Jean-Jacques Rousseau thought, humankind is born cruel, and it is only education that can make it 'good'. The basic values of a culture are those instilled very early, from age two to five years, and in general by the parents. In present times, the disappearance of the traditional family has been encouraged by an economy of plenty, where there is food for everyone even outside the traditional family lines. This is why the basic values that have made our Western civilization are in danger. Yet, these values have led us through the scientific and industrial revolution and have given us a prevailing position in the world.

What are these basic values? The notions of good and evil, the respect for human life, the importance of the individual, the human rights, the benefits of education for all, and the faith in progress. The fact that these basic values have not been taught *soon enough* explains the decline in quality of our schools and the rise in crime.

In America as well as in Europe, we try to form pluralistic societies that accept the coexistence of different cultures. This does not simplify the problem, because different cultures may have basic values that are in conflict. Is it possible to redefine basic values accepted by all, and can we restructure the education of *very young* children (from age two to five) in order to teach them common basic values? This is the immediate problem to be solved in the twenty-first century.

Of course, there are other immediate problems. The exponential growth of the world population must be stopped; the inequitable distribution of food and raw material is at risk of growing worse; pollution is also worsening. The danger of a runaway greenhouse effect exists because of our irresponsible production of CO_2 through the burning of fossil fuels (coal, oil and gas).

The fossil carbon was produced by the death of plants and animals over nearly 500 million years and we are going to burn it up within a few centuries (which is a million times faster than it was formed) in our electric power plants and our automobiles. This is a waste whose consequences cannot be assessed.

Dominating all these problems, stands the ghost of a nuclear or thermonuclear conflict (not to mention biological warfare), made more likely by the proliferation of nuclear weapons. Will we manage to prevent this proliferation? Will we reduce drastically the number of cultures that remain in open and often violent conflict? When we have solved the most urgent of these problems, we can then think about the next millennium. Cultures are the mental structures of *Homo sapiens*; their evolution and their selection should be able to work by the confrontation of ideas, without implying the destruction of individuals; but we are not there yet!

During the next centuries, one can easily imagine that we can colonize the Moon and Mars; perhaps we shall also build large space stations that are more and more self-reliant, where cities will be able to be born and develop. Depending at first on regular contacts with the Earth, they will probably be in orbit around the Sun, at nearly the same distance as the Earth; this would make it easier to keep the proper temperature and to capture the solar energy. One can then imagine similar cities with a propulsion system that would allow the population of a village (e.g. between 1000 and 3000 inhabitants) to leave the Solar System and establish a colony around a neighboring star, using this star as a source of energy. This could be compared with the way the Europeans of the sixteenth and seventeenth centuries colonized the Americas.

These pioneers will presumably have been preceded by machines that will have explored possible planets and helped to define colonizable locations. All this will take time, since in general a few hundred years will be needed for the journey. Unless hibernation techniques are used, we can assume that only the pioneers' great-grandchildren will reach their final goal. However, all of this falls within the limits of what we can begin within the next ten centuries.

Several interstellar propulsion systems have already moved beyond the dream stage. Some of them will soon be tested in space. Later, in order to reach nearby stars, they will accelerate during the first half of the trip, and decelerate during the second half, so that they will reach an average speed of 2% or 3% of the speed of light.

This limit of the thirtieth century that has just been considered is where most readers would find it comfortable to stop. However a single millennium is an extraordinarily short time span in the history of *Homo sapiens*. We mentioned that the average survival of a species is of the order of 8 to 10 million years. Over that immense time, without improving even the average speeds of interstellar flights, it would be possible to colonize, in successive waves, all possible planets in the Galaxy.

We do not claim that it will happen. We say only that we have the choice. For the first time in the history of life, our intelligence has given us the possibility of influencing the distant future by deliberate choices. We can even choose to survive much longer than 8 or 10 million years, which is the average survival time of an animal species. As a matter of fact, the awakening of intelligence has loaded the dice and completely changed the rules of the game. For instance, we are beginning to use molecular biology to our profit. At first, we will restrict ourselves to the correction of accidental genetic defects. Later, despite ethical, religious or cultural considerations, we will not wait for centuries before modifying the human genome deeper and deeper, in directions that we have not yet chosen, and with results wholly unforeseeable at the present time.

In our DNA, there is already a repair process that excises copying errors in the genome. By improving this process, it is conceivable that we could extend the survival of our species by diminishing the number of useless mutations; but would it be desirable? We might prefer the opposite: selecting rapid evolution of our genome by favorable mutations, leading to a kind of 'superman' that would be our heir and progeny. After all, we have been doing it blindly for a long time with cattle and horses! A deeper knowledge of the human genome may offer new techniques and unimaginable vistas.

There will be risks, and there will be moral, social, religious and cultural conflicts. However, our different cultures, which seem immutable in the short term, adapt very quickly to changes, and we are speaking in terms of thousands of years. There are still other possibilities that are difficult to ponder now, because we do not know enough about artificial intelligence. We can conceive of replacing dissipative structures based on carbon chemistry with new dissipative structures based on another chemistry, for instance that of silicon, and realize the dream of making organisms more intelligent than we are. They would in some sense be our descendants and our heirs, ready to leave Earth to go out to conquer the cosmos. Some might balk at the notion of being replaced by 'machines', so we should not venture too far in that direction.

Meanwhile, we can imagine that our human civilization will survive for a few million years. It will then probably have left the Earth in order to establish countless colonies in space. One of the principal reasons for such an expansion would probably be to increase the long-term chances of survival of the different cultural forms of our civilization, which would have spread throughout the Galaxy. A civilization confined to a single planet has very few chances of survival for the very long term; there are numerous possible cataclysms, such as a bombardment by another celestial body, a runaway glaciation or a runaway greenhouse effect, a cataclysmic change in the central star, which is the primary source of energy, not to mention cultural conflicts and wars of mutual extermination.

Inevitably, the civilizations would spread out, and some would sometimes forget their origins, as by the Middle Ages Europeans had forgotten the culture of ancient Greece. Some civilizations would decay and disappear, while others would continue to swarm indefinitely; eventually, some would end up by colonizing the whole Galaxy, in the sense that they would use the energy of all possible stars. In itself, the exploration and colonization of the Galaxy is the logical outcome of our exploration and colonization of the Earth that took place and was achieved during the last five centuries.

The distant future of life

Our Sun will burn for six more billion years, while increasing in brightness, first imperceptibly and very progressively. At a later time, it will make sense for a civilization to remain on Earth, only if it has taken the proper steps to secure its comfort, for example by modifying the greenhouse effect or by changing the cloud cover to increase the albedo of the Earth. Otherwise, it will not be too difficult to improve the living conditions of a Martian colony and to build large space stations further away in solar orbit.

Thereafter, the Sun will rapidly become a red giant, engulfing the orbit of Mercury, thereby vaporizing that planet. Measures to evacuate the Earth and Mars will eventually have been taken in time, at least if there remains a civilization to be taken care of.

At that time we can assume that the colonies that have been successful will have colonized practically the whole Galaxy. A number of readers will probably think that such anticipation belongs to science fiction; this is perhaps because they have not grasped the immensity of time that several million years represent. The staggering acceleration of technology during the twentieth century must be kept in mind. Less than 75 years separate the first faltering flight of an airplane from the walk of the astronauts on the Moon, and what is 75 years when compared to several million?

We still cannot foresee what techniques must be used to explore other galaxies, and this is perhaps a failure of our imagination. As the British science fiction writer Sir Arthur C. Clarke remarked, we may confuse magic with the capabilities of advanced civilizations. It is useless to discuss here the mathematical models which, for instance, suggest the feasibility of tunnels in space–time that are called 'worm holes'. At this stage, they remain pure fiction because they use general relativity without quantum mechanics, and they need a negative force of gravity to keep the worm hole open.

Let us rather consider the distant future, when our civilization is scattered throughout the Galaxy, and uses the energy of most of its stars. After some hundreds of billions of years, all the stars will have died out one by one. Inevitably, there comes a moment when the last of the very faint stars has faded out. If there are still civilizations, what are they going to do to survive, since their survival depends on the existence of temperature *differences*? Is life doomed to disappear?

We have often said that life depends on dissipative structures, which manage to evolve while staying always at some distance from thermal equilibrium. This remains possible as long as living organisms maintain throughout their bodies a flux of matter and energy that they draw from the environment. They are surrounded by a continuously rising entropy, a fraction of which they manage to reduce by extracting order, at the margins of growing chaos.

So long as the Universe continues to expand, it cools, and thus more and more entropy may become ordered. It seems therefore that life can exist indefinitely. Yet it must be recognized that technological problems will become difficult, because they require the use of temperature differences that become smaller and smaller. Will means be found to make use of the gravitational potentials around neutron stars or black holes, in order to draw off energy? It becomes impossible to make projections into such a distant future and still remain constructive; it is time to stop.

What we have just described for the distant future of human civilization is perhaps too optimistic. If the average lifetime of a civilization is so long, then other forms of life would have appeared elsewhere in the Galaxy, and a little before us, in terms of millions of years. Then, many extraterrestrial civilizations could have done as much as we will, and the traces of their colonies should be perceived in our neighborhood. Where are the other civilizations?

Where are the others?

If life has really emerged on innumerable planets, why have we not yet detected their presence around us? This question was asked by the famous Italiano-American physicist Enrico Fermi around 1950. More than ever, Fermi's question remains up to date because of our failure to detect any extraterrestrial signals, in spite of the recent efforts and progresses in this area (see Chapter 8).

According to Frank Drake's formula, if we are alone in the Galaxy, this is because the average lifetime of a technological civilization is less than 10 000 years; whereas, if this average lifetime were 100 million years, there would now be 100 000 coexistent civilizations, i.e. there would be one civilization per million stars of the right type. These civilizations would be scattered more or less 600 light-years apart. In the latter case, the most successful of these civilizations would already have colonized the whole Galaxy.

Why have we not been able to detect them? Brandon Carter remarked that, on the Earth, half of the life span of the Sun was needed on the main sequence for the emergence of intelligence. We have already mentioned the reason why: it took roughly 4 billion years to develop the eye and the brain. If this evolution bottleneck was overcome on Earth somewhat faster than average, most of the solar-type stars could have become red giants before the emergence of intelligence on their planets. Carter argued that we could be either the first, or at least among the first, civilizations in the Galaxy.

At the meeting about the Search for Extra-Terrestrial Intelligence (SETI) held at Santa Cruz, California, in August 1993, I mentioned a factor generally neglected in Frank Drake's formula: it is the need for the existence of giant planets in the cold zones of a planetary system, in order to deflect comets containing ices of water and volatile organic material, and cause them to bombard the rocky planets. Without this, the rocky planets could remain arid and never develop a biosphere. We have now discovered the existence of hot giant planets very close to their star, but we have no statistics yet on the existence of cold giant planets. If they are rarely found in the cold zones of planetary systems, this effect could diminish by 100 or 1000 the number of planets sheltering civilizations.

There are numerous other speculations on the reasons for the absence of contacts. For instance, the other civilizations are suspicious, or they are satisfied with waiting for our signals. They could already be aware of our existence, in which case they might be observing us, as one observes animals in a zoo. If there are many extraterrestrial civilizations, Drake's equation implies that practically all of them are much ahead of us. We are the candidates who want to 'join the club' and they may be careful before saying 'hello' and 'welcome'. They may wait until we have resolved our cultural quarrels and our nuclear wars.

They may also have a complete lack of curiosity. It is our curiosity that pushes us toward risk and exploration, and has led to the scientific and industrial revolution; but this is a new trend that only spread out in Europe from the Renaissance. Other cultures, like the ancient Chinese remained by choice frozen for a millennium. It cannot be ruled out that extraterrestrial civilizations could often have found analogous modes of cultural stabilization. Thus, we cannot yet reply to Enrico Fermi's question; we do not yet know whether we are alone.

Our exploration of the Universe has pushed back the frontiers of metaphysics. But metaphysics has not disappeared; as soon as we reach the limits of our knowledge, we still have to face the existence of the unknown. In order to conclude our world view, it is appropriate now to consider a few questions that may have been clarified by our quest, and to mention those which remain unanswered.

The answers of evolution

Evolution gives a direct answer to philosophical problems which, without its existence, have no obvious solution. A typical example is the question of knowing why we are not immortal. If we were immortal, evolution would not have been able to eliminate fast enough the overabundance of variations introduced by sexual reproduction. Unlike bacteria, which do not grow old, our cells are programmed to die after about 50 duplications. The price of the climb toward intelligence has been the loss of immortality. This was a corollary to the fast acceleration of evolution introduced by sexuality.

Now that we exist, there is a variant to the question: is there no way to become immortal? Could it be possible, in an indefinite future, not only to correct the effects of aging, but also to change the genetic instructions that require an inevitable death? The problem sets the needs of the individual against those of the human race. Evolution is not over; what would be the advantage of stopping the great experiment of the journey toward complexity, at a time when its final goal is not yet apparent?

Evolution also clarifies the problem of the existence of evil. The merciless elimination of the least fit is needed in each generation, to reduce to a reasonable level the survivors of the blind trials produced by new mutations. It is this constant pruning of the tree of life that has allowed its chief branches to grow up to their present height. As the visionary British poet William Blake said in 1810, in his 'A vision of the Last Judgment'

'To be an error and to be eliminated is a part of God's design.'

It is this constant pruning of errors that allows evolution to find untiringly the path toward progress, by a vast number of very small modifications, most of them having a very good probability of survival. Humankind has often believed that adversity is a curse, and that the struggle for life is a calamity, but both are at the heart of the process of evolution. Adversity gives the desire to struggle to change the situation, and without change no progress is possible.

Evolution of the brain induced the emergence of consciousness by a feedback from the immediate memory to the present moment, and also the awareness of the flow of time that derives from this feedback (without our being able to know, however, whether that fleeting flow of the present is an objective reality). The brain also creates feelings of pleasure and pain. These sensations come from that portion of the brain called the diencephalon, which contains the thalamus and hypothalamus and is now hidden by and under the cerebral hemispheres. The diencephalon is the part of the brain directly inherited from the reptiles. This 'reptilian' brain transmits its sensations to the huge cerebral hemispheres, which filter them and add an awareness of good and evil. Pleasure and pain are then interpreted in cultural terms, with variations in different cultures.

Humans have always tried to understand the origin of good and evil. In the Old Testament, Job clashed with the mystery of a just God who afflicts a no less just man with pain, misfortune and injustice. Yahweh refused to explain the mystery, and blamed Job for doubting His wisdom and His reasons.

We know now that adversity is needed in order to eliminate the least fit, and that the process works blindly. Pleasure and pain are only signs of confidence and of danger, green and red lights that natural selection has not ceased to reinforce, precisely because of their value for survival in the struggle against adversity. Their interpretation in terms of good and evil is cultural. In particular, it is only by a stretch of the imagination that we perceive the existence of the pain experienced by others, and that we interpret it in terms of good and evil.

The basic values of our Western civilization, which led to our belief in the importance of progress, also demand an interpretation of good and evil that implies respect for human life and the protection of human rights. These basic values are anchored in a tradition that has secured not only our survival, but also the success of our culture. In other societies, they can be very different; for instance, cruelty can be an accepted practice; tortures, amputations and death penalties are common punishments, and suffering among animals is ignored.

As for understanding why Yahweh refused to give Job an answer, it is probably because He gave up explaining that it did not make sense to interfere constantly in human affairs. The laws of the Universe had been set once and for all in order to urge life on to a constant ascent toward higher and higher peaks. Job would not have understood such an explanation.

Free will and quanta

By and large, humans believe that they possess free will; that is, the possibility of choosing a line of behavior through voluntary decisions that can only be influenced, but not constrained, by external causes. Until the beginning of the twentieth century, free will seemed to be in direct contradiction to the total determinism of natural laws. Could one extend determinism to moral choices? Were those choices already completely determined by preexisting causes? This would seem to imply that our free will is an illusion, and that all our acts have been blindly predetermined.

Quantum theory has rather changed the situation. It reintroduced uncertainty, and hence indeterminacy that manifests itself each time we induce what is called the 'collapse' of the wave function, by the observation of a quantum phenomenon. Since the visible world consists entirely of atoms, which are quantum objects, the whole Universe cannot be totally predetermined, but has a certain amount of freedom in its future evolution that is shown by the Heisenberg principle at the microscopic scale. Hence future events are not all predetermined but have only a certain probability of happening, which makes the more distant future hazier and hazier.

Since quantum phenomena exist in the atoms that make up the neurons in our brain, it makes sense to accept that they give us the possibility of exercising free will and influencing the outside world by deliberate choices.

We cannot discuss this in depth because philosophers of science have not yet reached an agreement on the matter of the 'collapse' of the wave function. Our purpose here is only to emphasize that free will does not seem any more to be in contradiction with the laws of nature; hence humans can influence the future by deliberate choices.

Why does the Universe exist?

Why is there something instead of nothing? As Leibnitz said three centuries ago, nothing happens without sufficient reason, and nothingness is simpler and easier than existence. We 'explained' the origin of the Big Bang by a quantum fluctuation, and showed that it seems possible to explain everything else through relationships from cause to effect. The primordial quantum fluctuation has no cause, no more than there is any cause producing the quantum fluctuation which triggers the emission of light of an atom, or the radioactive decay of an atomic nucleus; such is the usual but still surprising interpretation of quantum theory.

Our failure to understand does not stem from any special property of the Big Bang, but from our difficulty in interpreting the fundamental fuzziness of quantum mechanics, which implies that empty space has strange properties at the submicroscopic level. Our only hope for clarifying the situation lies in the expected fusion of quantum mechanics with gravitation, into a unified theory of 'quantum gravity' that might emerge in the twenty-first century.

In the meantime, some physicists have already tried to fuse some known properties of each theory in specific cases. The British physicist Stephen Hawking has been one of the pioneers in this direction, predicting on this basis a slow evaporation of the black holes. But he has also combined the uncertainty principle of quantum theory with a cosmological model based on Einstein's general theory of gravitation. His model suggests that, in the four dimensions of space–time, the Big Bang is not even a singularity; it looks rather (with one more dimension) like the north pole on a sphere; just like the direction north loses its meaning at the north pole, the flow of time appears at the Big Bang because before that, time behaves as an ordinary dimension of space (see Appendix C).

Another cosmological model had been offered by Wheeler, using the quantum interpretation that he proposed with Feynman. For them, the wave 'collapse' implies two waves travelling toward the past and toward the future. They chose a model that will collapse in the future and is described by a hypersphere in four dimensions, whose two poles represent the initial Big Bang and the final Big Crunch. At the moment of the final crunch, all waves that go backward in time are 'focussed' on the Big Bang and trigger it. This feedback loop is therefore capable of making the Universe without needing a cause for its beginning.

Needless to say, the question of existence versus nothingness cannot be resolved by a circular argument of this type, and the existence of a Creator moves the stumbling stone one step back only. It remains possible that all our difficulties stem from our psychological perception of the arrow of flowing time; we do not accept the absence of causality before the Big Bang, i.e. at the epoch when time was just another ordinary dimension which did not 'flow'.

Through the years, theorists often spoke about adding a certain number of dimensions to space–time, the goal being to explain the different types of symmetry that govern the elementary particles and their interactions, in the hope of finding the synthesis that will unify gravitation and quanta. It was suggested that these extra dimensions would remain undetectable, because they would be curved in upon themselves and would form loops much smaller than the atoms themselves. At the time of the Big Bang, space–time was also that small, and the initial 'bubble' had, for example, ten dimensions. The Big Bang would have enlarged only the four dimensions of space–time, while leaving the six others microscopic. Because these ideas have not yet led to anything constructive, I mention them here for the record only.

Thus, we do not yet know why the Universe exists.

'The feeling of the mysterious'

From my balcony, I like to look at the sky when the night comes. The stars appear one by one, and later in the night, the lights of the village go out at last; only the babbling of the creek is heard, and one smells the fragrance of pines. We are in our small dwelling at Vail, in the Rocky Mountains of Colorado, where the sky is very dark because of the altitude. In the calm of the night, I feel free from daily concerns, and sheltered from the muffled noises of civilization. Then I have the feeling of being in direct contact with the pure and silent Universe of the stars, and this is a nearly religious feeling.

Above the horizon, raggedly torn by the mountains, rise the constellations of Perseus and Andromeda. With the naked eye, I can finally just perceive the great galaxy of Andromeda. Its apparent size is larger than the Moon, but its light has travelled for more than 2 million years before reaching my eye. I ponder then about those innumerable galaxies much too far away to be seen except with the largest telescopes, and I wonder about the meaning of this Universe whose immensity I cannot comprehend.

However, this Universe has not always been so large, and it seems to have emerged from very little indeed. As time flowed, the Universe branched out into more and more ramifications that teemed and accelerated without pause. It moved through multiple stages, constantly climbing the steps toward a greater and greater complexity, based on cosmic processes present everywhere, so that the same evolution toward life must have taken place on a vast number of planets. It looks like an ascent that inevitably goes from matter to mind.

The Universe is still young, hence the ascent toward more complex entities with intelligences higher than ours may be only beginning. The summit of such an ascent is not yet discernible, so that it is impossible to guess whether it has a final purpose. But the whole scaffolding leading to life and intelligence is too well ordered, and it is tempting to believe that it has a hidden meaning. Up to the present, as the

American geologist Preston Cloud has prettily expressed it, 'stars have died so that we could live'.

We are indeed made of stardust, and it is quite recently that this stardust became aware of its own existence, and pondered why the Universe existed. We still know too little to answer the most important questions, but we are so young! We have just barely emerged in the Universe. However, we can make more and more deliberate choices, and take more and more significant decisions that will influence the future. If we really want it, our civilization will be able to escape the pitfalls of cultural conflicts and nuclear wars and set off into space, to the conquest of the Universe, in order perhaps to understand one day the reason for our existence.

For it is difficult not to ponder and wonder about the ultimate end of our ascent to intelligence, and the reason for all this undertaking. No philosophy has ever shed light on the profound reason for our existence; no religion has ever clarified the 'grand design of God' when He created the Universe. 'The most beautiful experience that one could have, is the feeling of the mysterious', wrote Einstein.

The night is quiet, the creek babbles, and the stars twinkle...

The ultimate mystery of the cosmos is its existence.

APPENDIX A

THE STANDARD MODEL OF
THE PHYSICS OF
ELEMENTARY PARTICLES

Matter is made up of two kinds of elementary particle, *quarks* and *leptons*, each having six varieties divided into two groups of three:

> *Quarks*
>> up, charmed, top (all three have charge $+2/3$)
>> down, strange, bottom (all three have charge $-1/3$)
>
> *Leptons*
>> electron, muon, tauon (all three have charge -1)
>> neutrino$_e$, neutrino$_\mu$, neutrino$_\tau$ (all three have charge 0).
>> The three neutrinos are not interchangeable; they are
>> distinguished by indices
>
> *Usual matter*
>> matter we are familiar with is made up only of:
>> the *up* quark, the *down* quark, and the electron;
>> in particular:
>>> the *proton* is made up of three quarks: *up*, *up*, *down*
>>> (charge $+1$),
>>> the *neutron* is made up of three quarks: *up*, *down*,
>>> *down* (charge 0).

Forces of nature

Four types of force of different symmetries occur among matter particles; these forces are carried by particles of a new kind:

(1) electromagnetism, carried by the *photon* (light, X-rays, γ rays, etc.)

(2) the weak force is carried by three particles: W^+, W^- and Z^0

(3) the strong force is carried by eight 'colors' of *gluons*

(4) gravitation is carried by the *graviton*

The standard model unifies electromagnetism and the weak force, as if the three particles W^+, W^- and Z^0 were actually heavy *photons*.

The existence of the *top quark* had been predicted for a score of years; it was finally observed in 1994, completing the list of elementary particles.

The only force-carrying particle that has not yet been observed is the *graviton*; however, indirect proof of its existence was established in 1984, following the discovery of a pulsar in a binary system. The period of its revolution around the companion star shortens by 75 milliseconds per year; hence the two stars are spiralling in toward each other whilst losing gravitational energy. This effect is predicted by the theory of general relativity, which explains it in terms of gravitational waves, transported by the *gravitons* of the standard model. For this discovery, the American astronomers Russel A. Hulse and Joseph A. Taylor received the Nobel prize for physics in 1993.

APPENDIX B

SYMMETRY IN PHYSICS

Symmetries are features which are preserved after a specified operation. For instance, two figures are mirror-symmetrical when we can turn one of them over the other and check by transparency that it coincides with the original. We call this property an *invariance* of the mirror reflection.

In this case, there is *invariance* in the drawing when it is turned over. Circular symmetry arises from the invariance of the length used as the circle radius. Symmetry of the equilateral triangle comes from the invariance of the length chosen for the three sides.

It is possible to extend the idea of symmetry to time. An invariance in time is called a 'conservation'. A mass that remains the same in the future as in the past comes from *mass conservation*, which expresses a time invariance, that is a symmetry when time flows. All conservation laws (of mass, energy, angular momentum, electric charge, etc.) are thus time symmetries.

An *antisymmetry* is also an invariance in absolute value, but with a change of sign. For example, a negative electric charge is antisymmetric to the same positive charge.

The symmetry properties of elementary particles, as well as those of the four distinct forces of nature, appear through particle interactions. These numerous symmetries are usually described as if they were geometrical symmetries in an abstract space of multiple dimensions.

A symmetry group is a set of properties that remain symmetric or antisymmetric over a specified type of operation. Symmetry groups are described in the abstract space mentioned above, which is in general a *phase space*; that is, a space where some dimensions mean physical properties such as mass, speed, momentum, spin, etc. This allows abstract symmetries to be described as if they were space symmetries.

They include in particular all conservation laws. The symmetry groups called $SU(5)$ are all the possible symmetry groups that correctly describe the set of all interactions between elementary particles, resulting from the four forces of nature.

APPENDIX C

THE STRANGE ROLE OF TIME IN RELATIVITY

In order to show the strange role of time in relativity, I will describe two examples that seem to contradict our intuitive feeling that the flow of time is different from a dimension in space. At this stage, let us remember that the two complementary theories called special and general relativity have been repeatedly checked by experiments and observations, and their consequences described hereafter are not in doubt.

In geometry, we all know that the length of the hypotenuse z of a right-angled triangle is connected to the length of the two sides x and y by the following relation:

$$z^2 = x^2 + y^2$$

In relativity the *interval r* separating two *events* is expressed in a very similar way; first, we will assume that the time span t and the spatial distance x are measured in the same units (for instance, years for time and light-years for space). In these units, the velocity of light $c = 1$. Then the interval is:

$$r^2 = x^2 - t^2$$

When in rapid motion with respect to one another, different observers of the two events see *only* the *interval r* as *invariant*, whereas x can become a different distance x' and t can become a different time lapse t'. That these distance and time spans should be relative is the essence of relativity, and we are no longer too surprised by it. The surprise comes from the minus sign before the square of time, instead of the plus sign in the formula for the square of the hypotenuse. Unfortunately, r and t cannot change roles, because r is invariant while x and t (or x' and t') are relative and depend on the observer's motion. Hence, there is no way to get rid of the minus sign.

It is this negative sign in front of the square of time that makes time's behavior different from that of space coordinates, and causes the geometry of space–time to be counter-intuitive.

As a first example, let us consider the flow of time near a black hole. We imagine an observer at a distance x from the center of a black hole with a Schwarzschild radius x_0. Let us call t_0 the proper time of the observer (that is, the time he reads on his watch). Since we are extremely far from the black hole, our time t is not the same. The two times t and t_0 run differently, according to the following relation:

$$t_0^2 = (1 - x_0/x)t^2$$

Very far from the black hole, x is very much greater than x_0, so that the fraction x_0/x is close to zero; the parenthesis is close to 1; and hence the two times are practically the same.

But in the vicinity of the black hole, the square of the proper time slows down in proportion to the parenthesis. For instance, at a spot x located at two radii x_0 from the black hole center, the parenthesis $(1 - \frac{1}{2}) = \frac{1}{2}$; there, the proper time of the observer runs $\sqrt{2}$ slower than ours.

The same computation shows that, at a distance $x = 1.01\, x_0$ (still outside the black hole, but very close to its surface) the proper time runs 10 times slower than our time. Finally, if the observer is right on the black hole surface, then the parenthesis becomes nil; whatever the time t, time t_0 remains the same by the observer's watch; seen from our vantage point, time has stopped at the surface of the black hole.

Surprises are only beginning. Within the black hole, x is smaller than x_0 hence x_0/x is *larger than 1* and the parenthesis becomes negative, meaning that the square of the proper time is a negative number. Within a black hole, the interval formula is still the same, but since t_0^2 is negative, if one uses its absolute value $|t_0^2|$, one has:

$$r^2 = x^2 + |t_0^2|$$

which shows that time now has spatial properties, since it respects the rule of the square of the hypotenuse, like every other spatial dimension!

We have just shown that the real distinction between time and space lies in the negative sign that comes before the square of time. There is nothing magic about that; it comes from the finite speed of transmission of light, but that topic would take us into too much detail.

Stephen Hawking believes that there was another situation where the square of time could lose its negative sign. This was right at the beginning of the Universe, when all extant dimensions were still small enough to be influenced by the fluctuations arising from the Heisenberg uncertainty principle. When a time span is short enough, a fluctuation of quantum origin can change the sign of the square of time, so that time properties disappear from space–time, and only a small bubble is left in four spatial dimensions. Extrapolating backward in time, this leads us to the primordial bubble, waiting for a quantum fluctuation that will release the flow of time and trigger the Big Bang.

Hawking's interpretation mixes relativity and a quantum principle, at a time when a general theory of quantum gravity does not yet exist. One will have to wait for that theory to know whether Hawking's intuition was well founded.

APPENDIX D

THE MEASUREMENT OF LONG TIME SPANS AND THE AGE OF THE UNIVERSE

To understand cosmic evolution, it was necessary first to evaluate the immense times involved. It began with geology. To find the age of a rock, one method came out on top: that of measuring the time elapsed from the moment when a radioactive element was confined in the rock. Uranium-238 (^{238}U) suits this particularly well, because it decays into lead-206 (^{206}Pb) with a half-life of 4.5 billion years. This half-life is the time needed for *half* of the radioactive substance to decay. After two half-lives, there is only ¼ left; after three half-lives, ⅛, etc. This is what is called an exponential decay.

The ratio of ^{238}U to ^{206}Pb present in a rock is a direct measure of the age of *solidification* of this rock. When a rock solidifies, the radioactive clock is reset to start at zero, because there is no ^{206}Pb in the uranium oxide crystal just formed (lead remains in the liquid state in the original magma or lava). The rate of radioactive decay is extraordinarily constant, and nothing short of destroying the rock can influence it. This stems from the fact that radioactive reactions call for much higher energies than do chemical reactions.

The oldest terrestrial rocks are 3.8 billion years old. NASA astronauts have brought back lunar rocks; the oldest of them are 4.1 billion years old. Most of the carbonaceous chondrites (coming from the asteroid belt) are all of the same age: 4.6 billion years to within 0.1 billion years. These are the most primitive meteorites known.

The *original* abundances of all radioactive elements must have been the same on the Earth, on the Moon, and in the meteorites, because they came from a common source. Their *extant* abundances imply that the branching from this common source of matter took place 4.56

billion years ago. This is assumed to be the time of the dust settling out from the gas in the solar accretion disk, which we used as the age of the Solar System. This fits in well with the age of the Sun, which is between 4.5 and 5 billion years; it cannot be assessed more accurately, because it depends on the internal structure of the Sun and on the change in its luminosity as it grows older.

Since all radioactive substances decay, but nevertheless still exist, they cannot have existed forever. They had to be made at a date that can be estimated from the ratio of their abundance to that of their decay products within the whole Galaxy. Although less accurate, the assessment depends also on the isotopes used, and fixes their *average* age at between 7 and 11 billion years. Since most radioactive elements are assumed to have been created in supernova explosions, their emergence was progressive and started some 11 billion years ago. This is roughly consistent with the ages determined for the oldest stars in the galactic disk. On the other hand, the Galaxy is enveloped by a halo of 'globular clusters' whose stars are older than those of the disk. The halo contains more than 100 globular clusters. Each cluster contains about 100 000 to 1 million stars, packed closely together. So narrow a space distribution implies that all these stars are at about the same age.

The only factor that strongly influences the time span of a star's evolution is its mass. Massive stars evolve hundreds of times faster than very low mass stars. For a particular globular cluster, the study of the masses of those stars that are in the process of becoming red giants is a good criterion for the age of the cluster. The oldest of these clusters were thought (until 1996) to be 14 billion years old, with an uncertainty of less than 2 billion years. The situation changed abruptly in 1997 (see later).

The expansion of the Universe was slow to be discovered because it ran counter to presumptions held since time immemorial. The speeds of recession of some 'spiral nebulas' (as they were called then) was mentioned first by Vesto Slipher in 1914, but it was not yet known that these objects were distant galaxies.

The recession velocity of distant objects is easily measured from the shift of all spectral lines of the observed object toward the red end of the spectrum. This red shift comes from the Doppler effect, which diminishes the apparent vibrational frequency of the light coming from the receding object.

But the measurement of distances of faraway galaxies has remained a major problem. In 1925, the American astronomer Edwin Hubble (whose name was given a half-century later to NASA's Space Telescope) started to evaluate the distance of faraway galaxies. Hubble found that their recession velocity v was nearly proportional to their distance d; the constant of proportionality H, obtained from the formula $v = Hd$, is still called the Hubble constant. The value of H found by Hubble was nearly 10 times too large, because the farthest distances were estimates that were 10 times too short. Soon revised to more reasonable values, the different estimates remained, however, a matter of controversy that would last for more than 60 years.

In February 1997, the new results of the satellite Hipparcos were announced. This satellite of the European Space Agency was designed to measure the parallaxes, hence the distances of 120 000 nearby stars, with an accuracy 100 times better than from the ground. Statistically, the distances found were 10% greater than expected. Since the parallaxes of nearby stars are the basic yardsticks for other, larger distances, the present results enlarge all distances known in the Universe. In particular, all globular clusters are 10% further away, hence their red giant stars are 21% brighter than was believed before. This implies that they burn 21% faster; hence the oldest globular clusters are not 14 but only 11 billion years old. The discrepancy brought about by the previously determined age of the oldest globular clusters has now virtually disappeared.

If the value of the Hubble constant had never changed in the distant past, it would now fix the distance of the 'event-horizon' at 12 billion light-years, hence the date of the Big Bang being set 12 billion years ago. However, there is no strong reason to believe that H has never changed in the past. Since in looking far out into space one is also looking back in time, in principle H could be measured for different epochs. Unfortunately, the measurements become less and less accurate at faraway distances, and become deceptive beyond 5 billion years ago. Some cosmological models claim that H was larger in the past, whereas some others claim that it was smaller. All that can be said at the end of the twentieth century is that there is no longer any serious contradiction coming from the observational data, and that the Universe is probably between 10 and 15 billion years old.

THE STANDARD MODEL OF
THE BIG BANG

The standard model does not try to explain the cause of the Big Bang. It starts from the present conditions of average density and temperature in the Universe. If we go backwards in time, the Universe was smaller; its density and temperature can be computed for some typical epochs in the past.

(1) *Average density.* The present Universe is very unhomogeneous, so that its average density can be estimated only by using a very large volume, for instance, a cube of 500 million light-years on each side, for which the total mass of millions of galaxies can be assessed. The average density found by this method is a little less than 10^{-30} g/cm^3.

(2) *Average temperature.* There are now still 3 billion photons at 2.7 K in the fossil radiation coming directly from the Big Bang, *for each hotter photon* arriving from the stars. These stellar photons are therefore the insignificant and negligible traces left by the fireworks from the primordial explosion. The *average* temperature of the Universe is 2.7 K.

(3) *Expansion velocity.* Its rate is given by the Hubble constant, H, which can be taken, for instance, as $H = 25$ km/second per million light-years.

In Table E.1, results have been rounded to the nearest factor of 10. The Universe cooled, as for example in an explosion gases cool as they fill a larger and larger volume, so that it is easy to compute the temperature and density at different times in the past. A last column has been added, where the average distance between extant particles is given. Only the result on the first row is deceptive, because the present

Table E.1. *Standard model for the expansion of the Universe (the table reads backwards in time)*

Milestone epochs in the history of the Universe	Age of the Universe (seconds)	Relative size of Universe	Temperature (K)	Mean density (g/cm^3)	Distance between particles (cm)
1. Present time	10^{18}	1	3	10^{-30}	100
2. Onset of transparency	10^{13}	10^{-3}	3000	10^{-21}	0.1
3. End of nucleosyntheses	100	10^{-9}	10^9	10^{-3}	10^{-7}
4. Onset of nucleosyntheses	10^{-2}	10^{-11}	10^{11}	10^3	10^{-9}
5. Quarks make hadrons	10^{-6}	10^{-13}	10^{13}	10^9	10^{-11}
6. Electroweak range	10^{-10}	10^{-15}	10^{15}	10^{15}	10^{-13}
7. Grand unification	10^{-34}	10^{-27}	10^{27}	10^{51}	10^{-25}
8. Quantum gravity	10^{-42}	10^{-31}	10^{31}	10^{63}	10^{-29}

Universe is very heterogeneous. If the Universe were formed by a homogeneous mist, the distance between individual atoms would be about 1 m. Beginning with row 2, the Universe was homogeneous: it represents the fireball conditions at the instant when it became transparent. The interval between the first two rows thus covers the whole history of the Universe, with the formation of galaxies and stars.

The Big Bang description begins at row 2, and goes backward in time. Rows 3 and 4 go back to the first hundredth of a second after the Big Bang, and their data are seen to correctly interpret the astronomical observations on the abundance of the light elements still present in the Universe.

Rows 5 and 6 are interpreted by the experiments with large particle-accelerators, which have explored the range of those very high temperatures where the undifferentiated quarks 'condense' into droplets of two and three quarks (the quark–antiquark droplets, called *mesons*, and the three quark droplets that make protons or neutrons). Since in row 6, the distances between quarks reach 1 femtometer (fm), it implies that, at that time, the Universe is a giant atomic nucleus where the strong and the electroweak forces prevail. We reach here the limit of validity of the standard model, because it does not take nuclear forces into account.

In order to delve into rows 7 and 8, the Unification of Forces must be considered; this seems a normal extrapolation, but such a theory is not yet either complete or verified. This theory leads to the *inflation*, described in Appendix F. Inflation makes the Universe explode like a gigantic radioactive nucleus.

APPENDIX F

THE CAUSE OF THE BIG BANG AND INFLATION

At the end of Chapter 1, it was mentioned that asymmetries in all the forces of nature disappear at extraordinary short distances, and that all force constants converge toward a single value.

Since the nuclear forces (strong and electroweak) are *confined* inside the atomic nucleus, the symmetry breaking of the forces must be produced by a phenomenon that takes place within the size of the nucleus (incidentally, this is what sets the nuclear size). The symmetry breakdown causes a *change of state*.

Changes of state, including familiar ones among solids, liquids and gases, imply a change in the symmetry properties. Ice forms crystals whose symmetry differs from that of water. Microscopic symmetries in the positions of atoms are not the same in liquid water and in steam. On the other hand, ice that turns into water absorbs heat without changing its temperature; it is the *latent heat* of the change of state. This latent heat arises from the entropy change coming from the symmetry change from ice to water. Moreover, while cooling, water often reaches a temperature below its freezing point without immediately solidifying: this is called *supercooling*.

Breaking of the original symmetry of all the forces has begun between rows 7 and 8 of Table E.1. About 10^{-35} seconds after the Big Bang, it can be assumed that the gravitational force had begun to fall in strength, followed by the decoupling of the strong nuclear force from the electroweak force.

This change of state was a phase transition that broke the perfect original symmetry; its latent heat unleashed an enormous amount of energy that heated the Universe while triggering an *exponential* expansion that has been named *inflation*.

Inflation is going to last for an extraordinarily short time, about 10^{-32} seconds; since the bubble doubles in size every 10^{-34} seconds, after 100 doublings, it has inflated by a factor of 10^{30} before slowing down to a constant expansion, which is no longer exponential. Thus, this single bubble created the whole visible Universe. The latent heat release was not instantaneous; this delay, similar to water super-cooling, lasted only 10^{-32} seconds. The energy released induced the emergence of a profusion of elementary particles, a plasma of quarks and electrons, that soon 'condensed' into droplets of two and three quarks, the hadrons. From row 6 upward, nothing has changed in the standard model of the Big Bang.

Thus, the inflation theory explains the cause of the Big Bang in terms of the destruction of the total symmetry of the original vacuum by a quantum fluctuation. It is based on the apparent need for a grand unified theory for the forces of nature, whose mathematical form has already been studied even if its physics is not yet fully understood. Moreover, the existence of quantum fluctuations is required by the present structure of quantum theory, which is no longer in doubt, since it has been proven by observations of numerous quantum phenomena. The fact that a quantum fluctuation has no deterministic cause has already been mentioned. This is one of the mysteries of quantum theory that we have to accept. It is based on the unusual properties of empty space at very small distances, which are used to explain the Big Bang, and are observed in the laboratory through the emergence of virtual particles out of nothingness (see Glossary: Quantum fluctuation, Empty space, and Virtual).

APPENDIX G

CHIRALITY

Chirality is the property of those molecules that can exist into two symmetrical forms corresponding to mirror reflections, but cannot be superimposed on each other by a mere rotation in space. Left-hand and right-hand gloves are an example of chirality. Chiral objects must be three-dimensional, since two symmetrical plane objects can always be superimposed by a reversal in space.

Many of the molecules used by life are chiral. However, when they exist in non-living matter, most of the time one half is in the right-hand form and the other half is in the left-hand form. This is what is called a *racemic* mixture. In contrast, life nearly always chooses only one of these two forms. For instance, all proteins consist of left-hand amino acids, whereas RNA and DNA are always built up from right-handed sugars. When a living organism dies and decays, thermal fluctuations change molecular shapes at random, so that, in the long run, there is *racemization*. Since the opposite process does not exist, a mechanism was needed to trigger the emergence of life by selecting preferentially one of the two chiral forms. The continuity of life then becomes only a mere copying process.

Was the choice random? Two forms of life of different chirality could have emerged. Left-handed proteins could have eliminated right-handed proteins by a random evolutionary process. This matter does not seem fundamental for elucidating the origins of life, because all biochemical processes depend on chemistry; that is, on the electromagnetic interaction which is mirror-symmetric. This would imply that the right-handed form has just as many chances of succeeding as the left-handed form.

However, the weak nuclear interaction (which is responsible for beta radioactivity) does not have mirror-symmetry. This is what physicists call the *parity non-conservation*. For instance, the electron freed by a radioactive nucleus has its spin vector parallel to its velocity vector. Spin is an angular rotational momentum; a mirror-reflection changes the rotation direction (hence its spin vector is upside down), whereas the electron velocity remains in the same direction. The weak nuclear interaction is therefore a *chiral* phenomenon, whereas electro-magnetism is not. However, electromagnetism and weak nuclear inter-action have now been merged into the so-called 'electroweak' interaction. The resultant asymmetry is extremely small. Could there be an amplification mechanism of this primordial asymmetry that could favor the left-handed amino acids and the right-handed sugars? This question has been discussed in the scientific literature for several decades, and has not yet led to an unambiguous answer.

However, J. Cronin and S. Pizzarello have announced (*Science*, 14 February 1997, p. 951) that they have discovered an excess of 7% and 9% of left-handed forms for several amino acids present in the Murchison meteorite. The results cannot be attributed to terrestrial contamination, since some of the amino acids are not used by life and do not exist in the biosphere. Because the Murchison meteorite is a car-bonaceous chondrite formed 4.5 billion years ago in the asteroid belt, the results show the existence of an asymmetric influence on organic chemical evolution before the origin of life. In our present state of ignorance, these results suggest that other forms of life must have the same chirality as ours.

GLOSSARY

ABSOLUTE ZERO: the lowest possible temperature, at which there is no heat energy left (0 K).

ALBEDO: fraction of the light reflected by a celestial body like a planet.

ALLELE: one of two analogous but not necessarily identical genes that exist at the same location on a pair of chromosomes. When two alleles are identical, the organism is homozygous for that gene; when they are different, the organism is heterozygous.

AMINO ACID: organic molecule with a radical $-NH_2$ and an acid function $-COOH$; the proteins are built up from 20 different types of amino acid.

ÅNGSTRÖM: unit of length equal to 10^{-4} micron (μm) or 10^{-10} meter.

ANGULAR MOMENTUM: measure of the rotary inertia of a massive object, around a specified rotation axis; the angular momentum remains constant as long as the mass is isolated from external forces.

ANTHROPIC PRINCIPLE: various forms of cosmological statements about the consequences of our existence on what we can observe in the rest of the Universe. The weak and the strong forms of the principle are discussed in the text.

ANTIMATTER: made of particles whose charge is antisymmetric to ordinary matter; for instance, the antielectron has a positive charge, whereas the electron has the same negative charge; all other properties remain the same.

ASTRONOMICAL UNIT (AU): average distance from the Earth to the Sun, used as a unit of length in the Solar System. Approximately 150 million kilometers or 8 light-minutes.

ATOMIC NUCLEUS: central region of the atom where nearly all its mass is concentrated. It is very small, of the order of 1 or 2 fm; it contains one or several protons, and often neutrons, bound together by the strong nuclear force, which prevents them from leaving the nucleus.

BETA RADIOACTIVITY: emission of an electron (and of a nearly undetectable neutrino) by the radioactive decay of a neutron into a proton; this decay is controlled by the weak nuclear force.

BIG BANG: popular name for the primordial explosion that would have begun the Universe and started its expansion.

BLACK HOLE: region of space–time that remains invisible to distant observers, because its gravity is so strong that nothing, not even light, can escape from it.

CELL: structural unit of living beings. The *prokaryotic* (without a nucleus) cells are the most primitive; they are found only in bacteria. *Eukaryotic* cells (with a nucleus) display a much more complex structure; they form unicellular microbes which are not bacteria (like the protists) and all the multicellular organisms (animals, plants and fungi).

CEPHEID STAR: class of red giant stars whose absolute luminosity can be estimated from the duration of their periodic brightness fluctuation. Their distance can then be deduced from a comparison of their absolute and apparent luminosities.

CHIRALITY: see Appendix G.

CHLOROPLAST: organelle ('little organ') present in the cells of plants; their pigment absorbs light and uses its energy to make photosynthesis, transforming CO_2 into food (sugars) with the release of oxygen into the atmosphere.

CHONDRITE: stony meteorite, coming from the asteroid belt (between Mars and Jupiter); since the formation of the Solar System, chondrites have not been differentiated, hence they are the most pristine sample available of the dust from which the terrestrial planets were made. They fall into four classes; namely two classes of *carbonaceous* chrondrites (defined as totally and not totally oxidized), *ordinary* chondrites (somewhat oxidized) and *enstatite* chondrites (totally reduced).

CORTEX: the cerebral cortex is the external part of the brain that is highly developed in humans; it encloses the diencephalon, which dates back from the reptiles. The cortex is the seat of conscious thought and intelligence, while the diencephalon is the seat of emotions.

COSMIC AFTERGLOW: see *Fossil radiation*.

COSMOLOGICAL CONSTANT: constant introduced by Einstein to avoid the collapse of the model of the Universe, based on general relativity. This constant gives the model an inner negative gravity which works against the gravity of the extant masses.

CRITICAL DENSITY: mean density of the Universe that would be just right to prevent it from falling back upon itself under its own weight. If the Universe had just the critical density, it would continue to expand indefinitely, although at a slower and slower pace.

DEUTERIUM: the heavy isotope of hydrogen, of atomic mass 2. Formed in the Big Bang, its amount has been stabilized by the rapid quenching that followed; it now 'burns' easily in the stars, where it is transformed into helium, like ordinary hydrogen, but at a lower temperaure.

DOPPLER SHIFT: apparent change in the frequency of light waves, caused by the relative motion of the observer and the light source. In astronomy, this shift is measured by the displacement of reference features in the spectrum.

ECLIPTIC: the great circle defining the apparent path of the Sun among the constellations; the plane of the ecliptic defines the plane of the Earth's orbit around the Sun; it does not coincide with, but is very close to, the mid-plane of the planetary system.

ELECTROWEAK: the electroweak force results from the merging of the electromagnetic force and the weak nuclear force, which become indistinguishable for energies around 100 GeV and at distances less than 1 fm (10^{-15} m); this corresponds to the size of the atomic nucleus.

EMPTY SPACE (OR 'VACUUM'): because of the existence of quantum fluctuations, the nature of empty space is not as simple as was thought before. The *average energy* of the quantum fluctuations in the vacuum is not nil. This energy seems to confer a transient (virtual) mass to all particles that transport nuclear forces, which constrains them to distances less than 1 fm (size of the atomic nucleus). These particles polarize space, which does not remain symmetrical in all directions. A totally symmetrical space ('false vacuum') exists only at high temperature.

ENDOTHERMIC: said of a chemical reaction that absorbs heat (in contrast to an exothermic reaction, which gives up heat).

ENTROPY: defines the amount of unusable energy; entropy can never diminish in a closed system; living organisms can diminish their entropy because they can reject the unusable energy to the outside world.

ENZYME: protein used as a catalyst for a biochemical reaction.

ERGOSTEROL: (also called provitamin D_2); steroid compound present beneath the skin, and converted into ergocalciferol (vitamin D_2) by ultraviolet light.

EUKARYOTE: see *Cell*.

EXON: part of the genome that is expressed in a particular living being; as opposed to *Intron*, a part which is not used, and often contains an incoherent message. The intron is not expressed in the individual, but is, however, transferred by heredity and remains present in the genome.

EXOSPHERE: outer part of the atmosphere where the gases are so rarefied that there are nearly no collisions between molecules, so that some can escape toward empty interplanetary space along ballistic trajectories. The exosphere temperature determines gas losses from the whole atmosphere.

EVENT: a point in space–time defined by its date in time and its location in space.

FEMTOMETER: unit of length equal to 10^{-15} meter. It happens to be the approximate size of the atomic nucleus; abbreviation fm.

FOSSIL RADIATION: also called microwave background radiation or cosmic afterglow: uniform radiation coming from all directions of the sky. Detected in mm and cm microwaves, it displays a temperature that is exactly the same at all wavelengths (about 2.7 K). It must be that of the Universe when it became transparent, 300 000 years after the Big Bang, but it was red-shifted by a factor of 1200 by the expansion of the Universe, so that its original temperature was between 3000 and 4000 K.

FRACTAL: configuration whose features display a geometric similarity that reappears at all dimensional scales. For instance, the size distribution of the lunar craters is a fractal, because their shapes and their irregularities are similar whatever the magnification; their structure has a scale invariance.

GALAXY: a system of several billions to hundreds of billions of stars; galaxies vary considerably in size and structure; most of them are not distributed uniformly in space, but arranged in clusters containing from a few up to 10 000 members. Most galaxies fall into two varieties: spiral and elliptical. Our Milky Way galaxy is a giant spiral. There are about 100 billion galaxies in the observable Universe.

GENOME: the whole of genetic information carried by DNA or RNA; in evolved organisms, the DNA is coiled around proteins to form chromosomes.

GLOBULAR CLUSTER: cluster of stars, spherically symmetrical, typically containing 1 million stars; 125 known clusters are scattered in a large halo surrounding our Galaxy.

GLUONS: the eight kinds of particles that carry the strong nuclear forces between the quarks inside the atomic nucleus; they stabilize the nucleus within their range; which is about 1 fm.

GRAVITON: particle that carries the gravitational force; not directly observed, its presence was demonstrated indirectly by the existence of gravitational waves that carry off energy, causing the orbital period of a double pulsar to diminish (Hulse and Taylor, 1993 Nobel prize for physics).

HADRONS: collective name for particles such as the neutron and the proton.

INFLATION: exponential swelling of the Universe during the first fraction of a microsecond after the Big Bang; although hypothetical at this stage, inflation seems to explain a series of properties of the Universe that otherwise have no obvious explanation (see Appendix F).

INTERFEROMETER: two telescopes looking simultaneously at the same object can be used as an interferometer; this means that their two beams of light can be combined, so that they *interfere* with each other, allowing a much sharper resolving power to be achieved, as from a much larger telescope.

INTRON: segment of the genetic message that is cut out at the time DNA is copied by RNA in complex living beings; the intron contains many errors and incoherent passages, owing to the hazard of mutations; see also *Exon*.

IONIZATION POTENTIAL: amount of energy required to remove an electron from an isolated atom or molecule; usually measured in electron-volts (eV).

ISOSTASY: in geophysics, the condition for the balance of the continents which, due to their smaller density, float upon the viscous mantle of the Earth.

ISOTOPE: one of two or more species of atoms of a chemical element; their nuclei have the same number of protons, but a different number of neutrons, so that their atomic mass is different.

KUIPER BELT: ring of comets on quasi-circular orbits beyond Neptune that is a remnant of the accretion disk where no planet has been formed.

K^0: neutral meson formed of two different varieties of quarks (one 'down' quark and one 'strange' antiquark) of equal and opposite electric charges, so the K^0 meson has no apparent charge.

KELVIN: temperature scale beginning at the absolute zero (-273 °C in the Centigrade scale); Kelvin degrees are identified by the symbol K; example: 300 K $=$ $+27$ °C. At very high temperatures, the difference between the two scales becomes insignificant.

MESONS: all varieties of particles formed by a quark and an antiquark (see Appendix A).

METALS: astronomers loosely call 'metals' all elements heavier than helium.

MICRON: also called a micrometer (μm); unit of length equal to one-thousandth of a millimeter.

MICROWAVE BACKGROUND RADIATION: see *Fossil radiation*.

MITOCHONDRION: organelle ('little organ') of the eukaryotic cell, which transforms nutrients into energy, generally by oxidation; it is believed that it is an ancient bacterium, introduced first as a parasite, then forming a symbiotic relationship with the infected cell. In particular, it possesses a different DNA that reproduces independently.

MITOSIS: division of the cell nucleus into two equal parts, in which all the chromosomes divide equally, leading finally to two cells identical to the first one. It is different from *Meiosis*, in which the chromosomes do not divide but are equally shared by the two new cells, which are then sex cells.

MUON: elementary particle which is a heavier electron (see Appendix A).

MUTATION: alteration of the genetic material of the cell; produced during the copying process, it can be compared to an accidental typing mistake. It is the source of the slow evolution of species.

NEUTRINO: elusive elementary particle of the lepton type; it is produced together with the electron when a neutron is being changed into a proton by beta radioactivity. Three different species of neutrinos exist, which are not interchangeable, but it is not known yet why they are different; it is not yet certain that they have a mass.

NEUTRON: neutral particle made of three quarks. It is one of the two components of the atomic nucleus, the other being the proton, which is positively charged.

NOVA: latin for 'new'; a nova is a star that suddenly becomes 100 000 times brighter, so that it looks like a *new* star in the sky.

NUCLEIC ACIDS: complex organic molecules composed of long chains of structural units called nucleotides; they comprise the genome of living cells. They are of two different types: DNA (deoxyribonucleic acid) and RNA (ribonucleic acid), which differ only in the nature of one of the four types of nucleic acid bases present in the nucleotides (uracil in RNA, thymine in DNA) and in the sugar present in the chain (ribose for RNA, deoxyribose for DNA). The DNA of human chromosomes occurs in the form of two twisted strands; the mRNA (messenger RNA) appears only in one strand. The rRNA (ribosomal RNA) is a structural part of the ribosome; finally, the tRNA (transfer RNA) transports the amino acids for assembly into a protein.

NUTATION: small periodic wobble of the Earth's axis; the Moon's attraction makes it wobble with a period close to 29 years.

OORT CLOUD: immense reservoir of comets almost spherical in shape, surrounding our planetary system. It was formed by ejection of icy planetesimals by the giant planets, during their growth by accretion.

ORGANELLE: one of the small organs present within the eukaryotic cell; the mitochondria and the chloroplasts are examples of organelles.

PEPTIDE: short chain of a few amino acids only, by contrast to the proteins which are very long chains of amino acids.

PERIODIC TABLE: table containing all the chemical elements, arranged in order of increasing atomic mass, and classified into seven horizontal rows reflecting the periodicity of their chemical and physical properties.

PHOTON: particle that carries the electromagnetic force and constitutes a quantum of light; the quantity of *action* (energy multiplied by time) carried by a photon is always the same (see *Plank's constant*) but its energy is inversely proportional to its wavelength.

PHOTOSPHERE: surface of the Sun, defined as the layer where it becomes opaque; in this layer, the Sun is sufficiently hot that its gases are ionized and become a plasma; the free electrons diffuse light and prevent the observations from penetrating any deeper.

PHYLOGENY: study of the evolution of a group of genetically related organisms: a phylogenetic tree is a sort of genealogical tree where the organisms are connected by the number of mutations that separate them.

PLANCK CONSTANT: this constant $h = 6.62 \times 10^{-27}$ ergs $\times$ seconds; it is the indivisible minimum of *action* called a *quantum*. Action is energy multiplied by time. For rotations, the Planck constant $h/2\pi$ is often preferred.

PLANCK LENGTH: this is a natural unit of length derived from the existence of the Planck constant. It is equal to about 10^{-35} m, which is extraordinarily small. If the Planck constant is h, the gravitational constant G, and the speed of light c, then the square root of hG/c^3 is a length which does not depend on the units chosen; this is the Planck length.

PLASMA: in physics, a plasma is the state of matter consisting of a mixture of ions and electrons. A plasma is similar to a gas except that it displays special magnetic properties. The Sun consists mostly of a plasma; the only familiar plasma is found in fluorescent tubes.

PRECESSION: the very slow, conical periodic motion of the Earth's axis, similar to that of a spinning top; the Moon's attraction makes the Earth precess with a period of about 26 000 years.

PROKARYOTE: earliest single-cell organisms that still survive as bacteria; prokaryotes do not possess a nucleus enclosed in a membrane, nor organelles, nor chromosomes with proteins and DNA. The blue-green algae are also prokaryotes. For this reason, they are not really algae, but are more properly described as cyanobacteria.

PROTEIN: long molecular chain, formed from about 100 to 500 amino acid building blocks. The amino acids of the proteins come in 20 varieties and their order produces the geometric structure of the protein. This is because the protein folds back upon itself in three dimensions, which makes it fit for a specific function, such as being a catalyst for some chemical reactions. Some proteins attach themselves to molecules which are not amino acids, but can give them different specific functions, for instance the pigment called porphyrin and an atom of iron enable *hemoglobin* to transport oxygen in blood; in chlorophyll, the iron is substituted by an atom of magnesium; this enables photosynthesis by absorption of solar light. The variety of the functions of all proteins is immense; they are both the architects and the construction materials of all organisms.

PROTISTS: unicellular microbes that are no longer bacteria but eukaryotes. Their single cell has a much larger degree of complexity than the prokaryotes. Protists are bigger than bacteria. Some of them are even visible with the naked eye.

PROTON: particle made of three quarks and charged positively; it is one of the two constituents of the atomic nucleus, the other being the neutron. The hydrogen nucleus is a single proton.

PULSAR: celestial source of radio waves that reach us in the form of regular pulses of extremely short duration. Pulsars are neutron stars that spin extremely fast and have an intense magnetic field; a pulsar acts as a rotating beacon.

PURINE: prebiotic organic molecule; its basic form contains two joined rings, one hexagonal and the other pentagonal. Its formula is $C_5N_4H_4$. It combines with a pyrimidine to make each rung of the double helix that makes DNA or RNA (see Figure 5.6). Adenine and guanine are the two nucleic bases that are purines.

PYRIMIDINE: prebiotic organic molecule, whose basic form contains a single hexagonal ring. Its formula is $C_4N_2H_4$. It combines with a purine to make each rung of the double helix of DNA or RNA. Cytosine and thymine are the two nucleic acid bases of DNA that are pyrimidines. In RNA, thymine is replaced by uracil, which is also a pyrimidine. The length of a unit of purine + pyrimidine remains exactly the same, whatever these two molecules are. This property allows an accurate geometric adjustment of the rungs of the double helix.

QUANTUM FLUCTUATION: spontaneous variation of the properties of empty space at extremely short distances invoked to explain the origin of the 'haziness' proper to quantum phenomena; in particular, a quantum fluctuation would explain the spontaneous emission of light by an atom, or the spontaneous decay of a radioactive nucleus (see also *Empty space*).

QUARK: fundamental elementary particle of the atomic nucleus, which is bound by the strong force; three quarks make either a *proton* or a *neutron*, whose combinations in varying numbers make the atomic nuclei. One quark plus one antiquark make one *meson*. Single quarks cannot exist outside the atomic nucleus, because the strong nuclear force confines them in a region of about 1 fm.

QUASAR: (contraction of quasi-star); intense celestial source of radio waves, whose position corresponds generally to a weak blue 'star', in whose spectrum all lines are (paradoxically) shifted strongly to the red. Because of their speed of recession, quasars appear to be the most distant objects known, often at a distance of 8 to 11 billion light-years. They are interpreted as produced by the very hot (blue) light of accretion disks surrounding massive black holes, probably at the center of the first galaxies.

RIBOSOMES: in the cell, numerous small granules composed partly of protein and partly of RNA. Their role is to make proteins from amino acids, using the instructions carried by the messenger RNA.

RIBOZYMES: different types of structurally distinct DNA or RNA that are enzymes; that is, specific catalysts of biochemical reactions.

SCALE INVARIANCE: geometric similarity that reappears at all magnifications, for all the features of a given configuration; it is said that such a configuration has the properties of a fractal.

SCHWARZSCHILD RADIUS: defines the radius of a black hole. Nothing can escape from inside this radius, not even a ray of light. When the mass M of a black hole is measured in solar mass units, its Schwarzschild radius R is given in kilometers by $R = 3M$. Thus, a black hole with four times the mass of the Sun has a Schwarzschild radius of 12 km.

SINGULARITY: a point where a mathematical function ceases to be defined, usually because it becomes infinite.

SPIN: intrinsic angular momentum of particles. Spin is a relativistic quantum property that cannot be wholly identified with a rotation; for instance, an electron must make *two* rotations to come back to its original position.

STELLAR POPULATIONS: astronomers divide stars into populations I and II. I have not mentioned it to avoid confusion, because population II comprises the oldest stars and population I stars are like the Sun.

STERADIAN: unit of solid-angle measure. The total solid angle about a point measures 4π steradiaus.

SUPERNOVA: final explosion of a massive star which, in a few days, becomes hundreds of billions of times brighter, so that it out-shines a whole galaxy for several months, before fading away; the explosion remnants remain visible in telescopes for millennia (example: the Crab Nebula).

SYMBIOSIS: permanent association of two organisms of different species who help each other to survive. For example, ruminants are in symbiosis with the bacteria in their digestive tract, which digest for them the cellulose in grass; in return, the bacteria are protected from the oxygen in the air.

SYMMETRY GROUPS: see Appendix B.

TECTONICS: in geology, plate tectonics is the study of the secular dis-placements and deformations of the large continental plates on the Earth's surface.

TELEOLOGY: explanation by some final purpose; it presumes a finality, a grand design which orients the Universe towards a goal. The strongest forms of the anthropic principle are a teleology, explaining the purpose of the Universe by our existence.

THIOESTER: class of chemical compounds analogous to the esters, but where oxygen is replaced by sulfur. The esters are formed by the reaction of a carboxylic acid with an alcohol, thus freeing a mole-cule of water.

TRANSCRIPTION: in biochemistry, copying the DNA by pairing the nuclear bases (a purine pairs with a pyrimidine and vice versa).

TRANSLATION: in biochemistry, passage from the DNA or RNA lan-guage (three 'letters' per 'word') to the vocabulary of the amino acids in order to form a protein.

URKARYOTE: one of the three ancestral branches of bacteria, which may have been the starting point of the symbiotic forms leading to the eukaryotes. The two other ancestral branches are the Archaeobacteria and the Eubacteria (or true bacteria).

VIRTUAL: used in physics to speak about a transient particle or prop-erty (like energy) that appears and disappears by virtue of a quantum fluctuation; for example, a virtual mass can appear only during the very short time allowed by the Heisenberg uncertainty principle. For very short distances, empty space shows the constant emergence and disappearance of virtual particles.

W^+, W^- AND Z^0: the three elementary particles that carry the weak nuclear force; this is the force that explains radioactive beta decay, as opposed to the strong nuclear force that is responsible for the stability of the atomic nucleus; the strong force is carried by eight different ('colored') gluons.

BIBLIOGRAPHY

The bibliographical references are aimed at the general reader who is curious enough to read some more about the different questions often very summarily sketched in the present book. These references have been classified by chapter. At the end are a few more general references to dig somewhat deeper into many of the problems mentioned.

CHAPTER 1. TO LOCATE HUMANKIND IN THE UNIVERSE
Morrison P. and Morrison P. H. 1982 *Powers of Ten*: Freeman, San Francisco, 159 pp. [About the relative sizes of things in the Universe.]

CHAPTER 2. THE RACE TOWARD COMPLEXITY
Barrow J. D. and Silk J. 1993 *The Left Hand of Creation: Origin and Evolution of the Universe*, updated. Oxford University Press, New York, 262 pp.
Hawking S. W. 1988 *A Brief History of Time*. Bantam, New York, 198 pp.
Longair M. S. 1991 *The Origins of Our Universe*. Cambridge University Press, New York, 128 pp.
Silk J. 1989 *The Big Bang*, revised. Freeman, San Francisco, 485 pp.
Weinberg S. 1977 *The First Three Minutes*. Basic Books, New York, 188 pp.

CHAPTER 3. THE STELLAR ALCHEMY OF METALS
Audouze J. and Vauclair S. 1980 *Introduction to Nuclear Astrophysics*. Reidel, Boston, 167 pp.
Shapiro S. L. and Teukolsky S. A. 1983 *Black Holes, White Dwarfs and Neutron Stars*. Wiley, New York, 645 pp.

CHAPTER 4. THE FORMATION OF THE PLANETS
McSween H. Y. 1993 *Stardust to Planets*, St Martin's Press, New York, 239 pp.

Taylor S. R. 1992 *Solar System Evolution*, Cambridge University Press, New York, 307 pp.

Morrison D. and Owen T. 1988 *The Planetary System*, Addison-Wesley, New York, 519 pp.

Murray B., Malin M. C. and Greeley R. 1981 *Earthlike Planets: Surfaces of Mercury, Venus, Earth, Moon, Mars*. Freeman, San Francisco, 385 pp.

Safronov V. S. 1972 *Evolution of the Protoplanetary Cloud and Formation of the Earth and the Planets*. Translated from the 1969 Russian edition, NASA TT F-677, US Dept. of Commerce, Springfield, VA, 206 pp.

CHAPTER 5. EMERGENCE OF LIFE

Chang S. 1982 *Prebiotic Organic Matter: Pathways for Synthesis*. pp. 261–280 in *Physics of the Earth and Planetary Interiors*. Elsevier Scientific, Amsterdam.

de Duve C. 1995 *Vital Dust*. Basic Books (Harper & Collins), New York, 362 pp.

Deamer D. and Fleischacker G. (eds.) 1994 *Origins of Life: The Central Concepts*. Jones and Bartlett, Boston, 431 pp. [Annotated collection of original papers.]

Hartman H., Lauten J. G. and Morrison P. 1986 *Search for the Universal Ancestors*. NASA PR-477, Washington DC, 129 pp.

Kauffman S. A. 1993 *The Origins of Order: Self-organization and Selection in Evolution*. Oxford University Press, New York, 709 pp.

Nicolis G. and Prigogine I. 1977 *Self-organization in Non-Equilibrium Systems: from Dissipative Structures to Order Through Fluctuations*. Wiley Interscience, John Wiley and Sons, New York, 491 pp.

Ponnamperuma C. (ed.) 1981 *Comets and the Origin of Life*. Reidel, Boston, 282 pp.

Thomas P. J., Chyba C. F. and McKay C. P. (eds.) 1996 *Comets and the Origin and Evolution of Life*. Springer-Verlag, New York, 296 pp.

Trân Thanh Vân J. *et al.* (eds.) 1992 *Frontiers of Life*. Editions Frontières, Gif-sur-Yvette, 499 pp.

CHAPTER 6. HISTORY OF LIFE

Bengtson S. (ed.) 1994 *Early Life on Earth*, Nobel Symposium no. 84. Columbia University Press, New York, 630 pp.

Cloud P. 1988 *Oasis in Space*. Norton, New York, 508 pp.

Emiliani E. 1992 *Planet Earth*. Cambridge University Press, New York, 718 pp.

Schopf, J. W. (ed.) 1983. *Earth's Earliest Biosphere: Its Origin and Evolution*. Princeton University Press, Princeton, NJ, 543 pp.

CHAPTER 7. AWAKENING OF INTELLIGENCE

Bronowski J. 1973 *The Ascent of Man*. Little, Brown & Co., Boston, 448 pp.

Gilling D. and Brightwell R. 1982 *The Human Brain*. Facts On File Publ. New York, 192 pp.

Ledyard-Stebbins G. 1982 *Darwin to DNA: Molecules to Humanity*. Freeman and Co., New York, 491 pp.

Smith C. G. 1985 *Ancestral Voices, Language and the Evolution of Human Consciousness*. Prentice-Hall, Inc., New Jersey, 178 pp.

Wills C. 1993 *The Runaway Brain*. Basic Books, New York, 358 pp.

CHAPTER 8. THE OTHER WORLDS

Baugher J. F. 1985 *On Civilized Stars: The Search for Intelligent Life on Other Stars*. Prentice Hall, New Jersey, 260 pp.

Baugher J. F. 1988 *The Space-Age Solar System*. Wiley, New York, 452 pp.

Cameron A. G. W. (ed.) 1963 *Interstellar Communication, a Collection of 1959–62 Reprints and Original Communications*. Benjamin, New York, 320 pp.

Heidmann J. and Klein M (eds.) 1991 *Bioastronomy: The Search for Extraterrestrial Life*. Springer-Verlag, New York, 413 pp.

Hoyle F. 1983 *The Intelligent Universe: A New View of Creation and Evolution*, pp. 83 and 86. Holt, Rhinehart and Winston, New York, 256 pp.

Shklovskii I. S. and Sagan C. 1966 *Intelligent Life in the Universe*. Holden-Day, San Francisco, 509 pp.

CHAPTER 9. PERSPECTIVES

Barrow J. D. and Tipler F. J. 1986 *The Anthropic Cosmological Principle*. Oxford University Press, New York, 706 pp.

Flood R. and Lockwood M. (eds.) 1986 *The Nature of Time*. Blackwell, Cambridge, MA, 187 pp.

Healey R. 1989 *Philosophy of Quantum Mechanics*. Cambridge University Press, New York, 270 pp.

REFERENCES TO DIG DEEPER

Davies P. 1989 (ed.) *The New Physics*. Cambridge University Press, New York, 516 pp. [Inflation, quantum gravity, grand unified theories, the new astrophysics, chaos, quarks, particle physics, etc.]

Maran S. P. (ed.) 1992 *The Astronomy and Astrophysics Encyclopedia*. Cambridge University Press and van Nostrand Reinhold, New York, 1002 pp.

OTHER GENERAL REFERENCES

Gribbin J. and Rees M. 1983 *Cosmic Coincidences*. Bantam, New York, 302 pp.

Reader J. 1986 *The Rise of Life; The First 3.5 Billion Years*. Knopf, New York, 192 pp. (in quarto, many color illustrations).

Davies P. and Brown J. R. (eds.) 1986 *The Ghost in the Atom*. Cambridge University Press, Cambridge, 157 pp.

Cornell J. (ed.) 1989 *Bubbles, Voids and Bumps in Time: The New Cosmology*. Cambridge University Press, New York, 190 pp.

Barrow J. D. 1992 *Theories of Everything: The Quest for the Ultimate Explanation*. Fawcett Columbine, New York, 302 pp.

FIGURE INDEX

TABLE INDEX

NAME INDEX

SUBJECT INDEX

water, liquid state 225–31
weak nuclear force 15
white dwarf star 47

Yucatán impact crater 190

zodiacal light 122